AF180689

ASSESSMENT REFORM IN SCIENCE

Science & Technology Education Library

VOLUME 31

SERIES EDITOR

William W. Cobern, *Western Michigan University, Kalamazoo, USA*

FOUNDING EDITOR

Ken Tobin, *University of Pennsylvania, Philadelphia, USA*

EDITORIAL BOARD

Henry Brown-Acquay, *University College of Education of Winneba, Ghana*
Mariona Espinet, *Universitat Autonoma de Barcelona, Spain*
Gurol Irzik, *Bogazici University, Istanbul, Turkey*
Olugbemiro Jegede, *The Open University, Hong Kong*
Reuven Lazarowitz, *Technion, Haifa, Israel*
Lilia Reyes Herrera, *Universidad Autónoma de Colombia, Bogota, Colombia*
Marrisa Rollnick, *College of Science, Johannesburg, South Africa*
Svein Sjøberg, *University of Oslo, Norway*
Hsiao-lin Tuan, *National Changhua University of Education, Taiwan*

SCOPE

The book series *Science & Technology Education Library* provides a publication forum for scholarship in science and technology education. It aims to publish innovative books which are at the forefront of the field. Monographs as well as collections of papers will be published.

The titles published in this series are listed at the end of this volume.

Assessment Reform in Science

Fairness and Fear

by

BENNY H.W. YUNG

The University of Hong Kong

Springer

A C.I.P. Catalogue record for this book is available from the Library of Congress.

ISBN-10 1-4020-3374-5 (HB)
ISBN-13 978-1-4020-3374-2 (HB)
ISBN-10 1-4020-3408-3 (e-book)
ISBN-13 978-1-4020-3408-4 (e-book)

Published by Springer,
P.O. Box 17, 3300 AA Dordrecht, The Netherlands.

www.springer.com

Printed on acid-free paper

All Rights Reserved
© 2006 Springer.
No part of this work may be reproduced, stored in a retrieval system, or transmitted in any form or by any means, electronic, mechanical, photocopying, microfilming, recording or otherwise, without written permission from the Publisher, with the exception of any material supplied specifically for the purpose of being entered and executed on a computer system, for exclusive use by the purchaser of the work.

Printed in the Netherlands.

Dedication

This book is dedicated to

all teachers

who are in pursuit of protecting students' interests

Contents

Preface

Fairness and Fear — I was startled at the subtitle of the book when I first heard it! Our series has some imaginative titles but none so evocative as this one. But "fear" and "fairness" capture much of teacher thinking about assessment. Indeed, teachers struggle to be fair with students, certainly concerned (and often fearful) of failures to be fair, and repercussions that teachers can face as a result. Then there is the fearful enterprise of imposed assessment regimes that many teachers face. Dr. Yung's book allows us to hear from the *teachers* on these concerns — this is no top-down policy statement. His account is readable and highly instructive, and I hope that copies will find their way to the desks of many school administrators and policy personnel. We are indeed pleased to release this volume.

William W. Cobern
Book Series Editor (2000-2005)

Foreword by Peter Fensham

Two things are very clear to me about formal educational systems.

- *What is assessed in these systems determines what teachers and students recognize as knowledge of worth.*
- *Teachers in general are conscientious in doing their best to ensure that their students will learn this knowledge of worth well.*

Science has now been widely acknowledged to be a core or key subject in the overall curriculum of schooling at all levels. This high status for science is a recognition of the pervasive influence of science in 21st century societies and the consequent need to equip all students with an appropriate understanding of these influences.

Science is a great human endeavour, but it is a complex one. Interested men and women, for hundreds of years, have asked questions about the natural world in ways that slowly have led to what we today know as Science - a great corpus of knowledge and a variety of procedures for expanding this knowledge. In doing so, they have uncovered facts about natural phenomena, developed ideas and explanations about them, and invented products in the form of technologies that transform human lives and societies. And indeed, how we think about ourselves and the world itself.

If the complexity of science is taken seriously in its teaching in school there is immediately a need for methods of assessing learning that recognize the very different types of learning this subject includes. The first step in this search for authentic assessment was the recognition that the abilities involved in carrying out standard laboratory experiments and more open-ended practical investigations, could not be satisfactorily assessed by an external written test or by an externally organised practical examination. To not assess, would mean the down grading of this important aspect of science in the eyes of students and teachers. As Black (2004) has forcefully stated "if you wish to know how students are doing in practical science investigations, there are no surrogates. The only valid source of this assessment is the students' teachers, whether they like it or not".

The situation in Hong Kong on which this book is based is just such a case – the internal assessment of biology laboratory work by senior secondary teachers. Like good teachers everywhere, these teachers readily accepted their professional responsibility to assist the learning of their students. They understood and accepted that this included the assessment of their students' learning in a formative sense in order to provide their students with helpful information about what learning is intended, and to provide themselves with feedback about their teaching.

The role of assessing their students' practical learning in a summative sense was, however, quite a different matter. This was especially so when it did, as in this case, have major implications for their students' futures. Teachers immediately saw and experienced this new task as involving a conflict of interests.

How could they fulfill this role and at the same time be the helpmate of their students?

The task of validly assessing the practical aspects of science teaching is just the beginning of the similar demands for assessment by teachers as we continue to understand what teaching for scientific literacy includes. There is already awareness that affective outcomes of learning science should be outcomes of science education along with the traditional cognitive and practical ones. It seems that these new outcomes, along with the learning of creativity and intuition in science, and of socio-scientific decision making, will all be best assessed by students' own classroom science teacher.

This book, thus, comes at a most opportune time for science teacher educators. They are now charged, in their work with pre-service teachers and with the in-service of existing teachers, with assisting them to cope with these new role responsibilities. The case stories of ten teachers faced with the tensions of internal assessment are rich accounts of the different ways these responsibilities were tackled. Of particular help to teacher educators are the constructs of *fairness* and *professionalism* that the author developed from these stories of teachers in the acts of their practice.

Teachers will also find the book's stories to be mirrors in which they can find some images of themselves as they face the new challenges of teaching for scientific literacy.

Peter J. Fensham
Monash University

Foreword by Derek Hodson

In a career as science teacher and science teacher educator spanning almost 40 years, I have read numerous accounts of research projects and dozens of books in which self-styled experts theorize about educational practice and proffer advice to novice and experienced teachers alike. Most follow a fairly standard and entirely predictable format. In consequence, most leave little lasting impression and do little to change the practices of those whom they seek to influence. Once in a while, a book comes along that takes a refreshingly different approach. Benny Yung's book is one of those rare instances: a book that takes us into the day-to-day world of teachers with no preconceived framework to tell us what we should expect, how we should teach, or even how we should recognize good and less than good classroom practice. Instead, we are immersed in the daily round of curriculum decision making of a group of ten teachers as they struggle to implement a major assessment reform in Hong Kong schools. Only when readers are thoroughly immersed in the moment-by-moment concerns of these teachers as they deal with classroom incidents does the author reveal his own position on these matters. At which point, we realize that he has gently and subtly shifted our thinking in the direction he intended by means of a careful and systematic introduction of particular theoretical perspectives to make sense of the episodes we encounter. By reversing the usual flow of information, Benny Yung has produced a powerful text that, unlike more conventional texts, *will* influence teachers and teacher educators.

The educational context is the shift in examinations procedures for A-level biology (school leaving examination for 17-18 year olds) from a one-off external practical examination to a school-based continuous assessment approach – the Teacher Assessment Scheme (TAS). As might be anticipated, many teachers struggled with the issue of *fairness* and what it means in the context of a situation in which they are cast in the dual role of teacher and assessor. Inevitably, questions of when or if to intervene, and when or how to give advice, created tensions and pressures that led teachers to solve their dilemmas in substantially different ways. Dr Yung's discussion of three very different views of fairness resulting from teachers' deliberations provides an ideal vehicle for both pre-service and in-service teacher educators to deal with assessment issues. But it goes much further than that because it provides case studies that graphically illustrate how day-to-day teaching is, above all, characterized by constant and far-reaching decision making – a theme that Dr Yung is able to address in the subsequent chapters at gradually increasing levels of theoretical sophistication.

While all ten teachers were to some extent constrained by the regulations of the TAS, some were able to interpret them in ways that were consistent with their educational values and some were not. Clearly, some teachers were able to exercise control of their teaching by adopting a critical stance towards the mandated policy change, while others were unable to do so. The book is a fascinating insight into how these differences occur and, by extension, what teacher educators can do to enhance the possibility that teachers will retain or develop classroom autonomy. The book tells us very clearly that teachers don't just deliver the curriculum. They interpret it, modify it, develop it, refine it and re-interpret it. The book tells us very clearly that meaning does not reside entirely in curriculum documents. Rather, it is *negotiated* in the classroom, as policy documents are confronted by and scrutinized by a complex of teacher knowledge, beliefs, attitudes, aspirations, values and sense of professionalism. As Dr Yung points out, recognizing this reality is key to successful curriculum innovation. Quite simply, attempts at reform are doomed if they fail to take account of the day-to-day lives, needs and interests of the teachers whose practice they seek to change. What is needed, therefore, is substantial investment in professional development of teachers – a message that some education authorities have yet to heed.

Of course, enhancing the quality of teacher decision-making on curriculum matters entails not just enrichment of the intellectual resources on which teachers draw to inform their decision making, but also a fundamental reorientation of teachers' views of their professional responsibilities. While some might argue that to say this is not to say anything radically new, Dr Yung has gone way beyond mere rhetoric and has provided would-be innovators with a powerful *means* of engaging teachers in discussion of such matters. This is a book that can play a substantial role in bringing about professional development and addressing the question of what constitutes teacher professional consciousness. Its principal strength is located in the series of case studies of real teachers in real classrooms struggling with day-to-day dilemmas of how to conduct the TAS. Through Dr Yung's skilful portrayal of classroom life, we get to know the teachers and we come to care about them as they engage in the kind of classroom situations with which we can readily empathize. It is the immediacy of the classroom descriptions and analyses that gives the book its unique power; it is the richness of the case studies that prepares the ground for Dr Yung to lead the reader through a series of sophisticated analyses and interpretations of the classroom events. This is not to say that the book neglects theory in pursuit of description. There is an extensive appendix that provides a critical review of three bodies of theory relevant to the focus of the book: the goals and effectiveness of practical work in school science, issues of assessment and evaluation in science education, and the role of teacher beliefs and teacher pedagogical thinking in determining classroom actions. A second appendix deals at considerable length with the research methodology issues pertinent to the study. These two appendices constitute valuable review material for teachers, teacher educators, researchers and graduate students. It almost goes without saying that the structure of the book enables it to be used in variety of ways, depending on the needs, interests and predilections of the reader.

Although the book is located in the specific context of Hong Kong and the specific context of assessment of practical work in biology, the findings and insights of Dr Yung's research transcend both contexts and have much to tell us about the ways in which teachers can, and should, respond to mandated changes in any subject area, in any educational jurisdiction. Used appropriately, Dr Yung's book has the capacity to impress on teachers that their professionalism resides in the capacity to interpret curriculum documents and translate them into worthwhile and effective classroom practice through sound judgments taken in pursuit of student learning. The capacity to make such judgments is, of course, dependent on having sufficient knowledge to properly inform those judgments and sufficient confidence to make them. Professional confidence, in turn, is determined by the ways in which teachers view their role – that is, dependent on their professional consciousness. Dr Yung's book provides a treasure house of 'ordinary episodes' that teacher educators can use to enhance both knowledge and confidence, and can use to stimulate teachers to explore the notion of professional consciousness. It is many years since Lawrence Stenhouse asserted that there can be no effective and worthwhile curriculum development without substantial teacher development. I am delighted to write this foreword to a book that can do much to bring about the kind of teacher development that Stenhouse envisaged.

Derek Hodson
Ontario Institute for Studies in Education of the University of Toronto

Chapter 1

INTRODUCTION

Assessment, in whatever form it takes, is widely recognized as one of the main determinants of educational practice. Over the past few years, new approaches to assessment have emerged in a number of countries. These have come primarily from a variety of overlapping debates concerning the purposes and methods of assessment, and their impact on the process of teaching and learning. Some of these debates have taken place in the context of intense political debate about the accountability of education services. Others have been more concerned with how changes in assessment practices and procedures can contribute to improving teaching and learning. In either case, teachers' beliefs, knowledge, and skills are pivotal in bringing about change in assessment practices. In consequence, teachers' knowledge and beliefs should be a primary area for research (Shepard, 2000).

This book addresses the issues for teaching and learning that emerge when school-based assessment is mandated, its contribution to teaching and learning, and how teachers' beliefs and teacher professionalism become involved. The substance of the book is derived from a study of ten teachers in schools in Hong Kong, as they tried to change their practice following a reform of the assessment system for senior secondary biology. The study that led to this book was situated in the particular policy and curriculum context of schools in Hong Kong, and its ten case studies were located in this specific educational setting. Nevertheless, its findings do transcend the specificity of the setting and have wide implications in the educational contexts of many other countries. The rich stories of teachers in their professional actions, their struggles, worries, concerns, their visions and inspirations, as well as their professional learning, reported in this book, may provide a useful source of curriculum materials for professional development courses for teachers, not only in the areas of science education and school-based assessment, but also in courses that aim at developing

B.H.W. Yung, (ed.), Assessment Reform in Science, 1-9.
© 2006 *Springer. Printed in the Netherlands.*

teachers' philosophies of education more generally. These accounts of the teachers can be helpful to all teachers, whether in training or experienced, in the following ways:

- as a set of ideas to be debated upon and to act as a springboard for reflection on the purposes of assessment in education and on the role of teachers in these purposes
- as examples of practice that can be compared to the teachers' own existing practice; and
- as a source of models of practice for teachers to apply and test in their own classroom.

In sum, this is a book for those who care about thoughtful teaching, a book for those who believe that the teacher is the ultimate key to educational change and school improvement. Anyone who has taught school, or spent time talking with teachers will recognize, sometimes with uncomfortable clarity, the problems and issues described in this book. Those who are less familiar with the professional side of teaching are bound to find in this book that teaching is far more complex than they previously thought.

1. BACKGROUND OF THE STUDY AND ITS UNDERLYING RATIONALE

1.1 Assessment Reforms

The presumed centrality of students' practical work in the science curriculum, reinforced by a variety of curriculum initiatives beginning in the 1960s through to the recent renewed interest in learning science as inquiry, has raised a number of questions about how such work might best be assessed and accredited (Caillods et al., 1996). This is especially true in contexts, like Hong Kong, where assessments involve such high-stakes. Often, the conventional one-off external practical examination has come to be seen as seriously and manifestly defective. For example, in the UK, teacher assessment occurs for practical work in the Key Stage tests, in the General Certificate of Secondary Education (GCSE) and at the Advanced-level (A-level) examinations, and is relevant to this discussion.

The Hong Kong educational system is characterized as an examination-led system, where what goes on in the classroom is largely dictated by what happens in the public examination halls (see also 2.1). Though the competition for tertiary places has lessened in recent years, the emphasis on examination for selection purposes is still much stronger than in some other places. Indeed, it has been emphasized that any change in the Hong Kong educational system must first involve a change in the examination system

(Morris, 1990). Even the public examinations administering body itself is very conscious of the fact that "if it is not examined, it won't be taught", and "much thought has been given to how the examination process can be used to bring about positive and constructive change to the system" (Hong Kong Examinations Authority [HKEA], 1994, p.80). As an outcome, the former Hong Kong Advanced Level (HKAL) Biology Practical Examinations was recently replaced by a mandated school-based continuous assessment scheme – the Teacher Assessment Scheme (TAS), which is the specific context of this book. Advocates of the TAS have claimed that its implementation would have a liberating influence on the curriculum and would bring about a host of desirable curricular and pedagogical changes (Pang, 1992). Many of these issues are also pertinent to countries, like the USA, where assessment reforms are currently being contemplated.

Top down directives often take insufficient account of teacher knowledge and beliefs, and insufficient account of the realities of individual classrooms. Drawing on evidence from the UK, Torrance (1995a) was critical of a simplistic and mechanistic view of the implementation process and warned against the danger of such an optimistic view of the new assessment reforms. In the USA, Khattri, et al., (1998) reviewed the diversity of experiences of educators at 16 school sites and noted that "new approaches to student assessment alone are not sufficient to improve teaching and learning. Rather, the principles and ideas underpinning assessment reform must be clearly defined and understood at all levels" (p.144). In particular, they pointed to the importance of shifting the focus of teacher professional development from communication of facts to capacity building. But, what exactly does capacity building mean? For some reformers, improving teaching is mainly a matter of developing better teaching methods and, in this case, assessment skills. These things are important, but we are also increasingly coming to understand that developing teachers and improving their teaching involves more than giving them 'new tricks'. The way teachers teach is not merely a consequence of the skills they have acquired; it is also grounded in what they think and what they believe.

1.2 Teacher Beliefs and their Practice

There is a growing body of evidence supporting the premise that teachers do have theories and belief systems which play an important part in their cognition and behaviour in teaching (e.g. Fischler, 1994; Blake Jr., 2002). In numerous studies of how teachers implemented new initiatives in the curriculum (e.g. McRobbie and Tobin, 1995; Tobin and LaMaster, 1995) it has been found that when the philosophy of the curricular innovation is significantly different from the theories and beliefs held by the teachers, the challenge or demand on the teachers requires them to restructure their beliefs or to "domesticate" the curriculum in order to fit it into their belief system. Sometimes, such re-structuring or domestication may be problematic or even "personally threatening". Is this the case for teachers in the TAS? The TAS demands a reformulation of the teacher's role as both an assessor and a teacher. Additionally, underpinning the TAS is a radical shift in the philosophy of assessment and its role in relation to learning. These changes entail a shift in thinking which is referred to by Gipps (1994) as a paradigm shift, from a "testing" model to a broader model of "educational assessment". Whether these new beliefs can be translated into pedagogical form will depend on the personal beliefs of teachers and their willingness to change. This leads us to the focus of this book.

1.3 Teacher Professionalism and Educational Reforms

One of the assumptions underlying centrally directed educational reforms is that teachers, where necessary, will be both willing and able to adapt their teaching in appropriate directions. Yet there is considerable evidence to suggest that this is not so. Teachers mediate the external pressures upon them through the 'filter' of their own professionalism. Indeed, the TAS reform was understood by Hong Kong teachers in several very different ways. Some teachers perceived the TAS as creating constraints for their teaching while others perceived it as creating opportunities for their own professional development and subsequently heightening their sense of professionalism. The latter view was, of course, what the advocates of the reform hoped to encourage (Pang, 1992). They argued that the reform represented decentralisation of the power of assessment from the public examining body to teachers, entrusting them with more responsibility. So, there was a supposition about the positive effect of the reform on teachers' professionalism. But the questions remain: why did different teachers perceive the reform differently; on what did the teachers base their interpretation of the TAS regulations and their enactment of these in the classroom; and was this related to their sense of professionalism?

At the outset of the study reported in this book, one broad and four sub-level questions were posed. The broad question was: What relationships exist among (i) secondary science teachers' classroom practice, (ii) secondary science teachers' beliefs about science, teaching and learning, and (iii) their understanding of the school-based assessment reform in practical science?

The sub-level questions were:

1. What are the characteristic features of secondary science teachers' classroom actions in the TAS?
2. In the context of classroom teaching, what personal understanding / perceptions do secondary science teachers have of the TAS?
3. What are secondary science teachers' beliefs about science, teaching and learning?
4. What are the relationships among teachers' beliefs about science, teaching and learning, their understanding of the TAS reform, and the ways they implement the reform inside their classroom?

However, as the data emerged from the study, it appeared that the varying practices of the teachers, and the different ways in which the TAS regulations were interpreted by the teachers, could well be explained through the lens of teacher professionalism. Analysing the data in such a way has enabled me to gain a deeper understanding of the relationship between teachers' beliefs, their practice and their sense of professionalism.

2. METHODOLOGICAL APPROACH OF THE STUDY

What follows is a brief description of the methodologies employed in this study. Appendix B provides a more detailed discussion of the study design and the methods employed.

2.1 Data sources

The study comprised more than 50 classroom observations of ten biology teachers. For each teacher, at least four different practicals (laboratory teaching sessions) were observed, each lasting from two to four hours. Other lessons associated with the practicals concerned were also observed. These included pre- and post-laboratory sessions if they were implemented separately from the laboratory session proper. All the lessons were recorded via a wireless microphone attached to the teacher. As the researcher, I sat in a corner at the back of the classroom monitoring the recordings as well as jotting down notes of the events taking place, notes about the teacher, about students, and anything that struck me as significant.

The classroom observations were supplemented by post-lesson interviews that probed the teachers' pedagogical decisions as well as their associated thinking and beliefs. Several other interviews were subsequently conducted with each of the teachers to obtain a wider picture of their views on quality practical work teaching, assessment, teaching and learning as well as their experiences as a student and as a teacher. These sources provided in-depth data about the personal background of the teachers and the context in which the present study is embedded.

2.2 Data treatment and analysis

To interpret the data I used a method similar to that described by Erickson (1986), namely, a focus on "the immediate and local meanings of actions, as defined from the actors' point of view" (p.119). My primary aim was to find out, from the perspectives of the actors - the ten teachers - what was happening in their classrooms and why they acted in certain ways. This was done by reading the interview transcripts in relation to the classroom data to ascertain what the teachers saw as significant in their classroom practice.

To ensure that the study would yield outcomes that were authentic, trustworthy, credible and robust, the data collected, including transcripts of interviews with the teachers and my interpretations of the data, were shared with the teachers so that they could review the data and interpretations, agree or disagree with the assertions of the research, and suggest corrections, elaborations and summary statements.

Narrative accounts in the form of case reports have been used to convey the context of this study and the knowledge that was implicit in the stories of the teachers, and of myself as a non-participant observer. In this way, I felt I could emphasize the voices of the teachers in a credible manner, and emphasize my interpretations in the latter parts of the book.

2.3 Narrative accounts and interpretations

To assist readers in developing a sense of the whole teaching context against which the present analysis of the relationships between the teachers' practice and their associated beliefs is set, each teacher's story is prefaced by a portrait of the teacher concerned, followed by a narrative account of the teacher's practice in the TAS, together with the explanations offered by the teacher on his/her own practice and my interpretations.

3. ORGANIZATION OF THE BOOK

I share with many others, the view that teachers are the deciding factors in the success of assessment reforms. The organization of this book reflects my commitment to contributing to teacher professional development in this area. To cater for practising and prospective teachers, the structure of the book is organized around the case stories of the teachers. These are followed by a discussion of a number of issues that arise from this group of teachers' professed beliefs and their deployment in classroom practices. The relevant literature, theoretical underpinnings, and the research methodology that led to the case stories are included as appendices.

Chapter 2 provides the reader with more information about the context of the study on which this book is based – the HKAL Biology TAS – including the history of its development, its regulations and some prior studies on it. This aims to acquaint readers with sufficient background knowledge to appreciate what is happening inside the TAS classrooms.

Chapter 3 introduces the reader to a number of different foci and types of teacher-student interaction in the TAS classrooms. Each is illustrated by an actual example. The overall picture of these interactions is presented in terms of their relative proportions among the whole group of ten teachers. These quantitative results also provide the basis for separating out the teachers' different profiles of teaching and exploring how the teachers' beliefs connect with their classroom actions.

Chapters 4 to 11 examine, in depth, the classroom practices of eight selected teachers, in relation to their underlying beliefs, in a more focused and qualitative manner. The findings are reported in the form of individual case stories, one for each chapter. Each teacher's story is divided into three parts. The first part begins with a brief description of the teacher's personal, educational and professional background, followed by the teacher's vision of the goal(s) of education in general and science education in particular. This is followed by the teacher's espoused beliefs about (1) the nature of science, (2) the role of practical work in science education, and (3) teaching and learning in general. The second part of the teacher's story reports on the characteristic features of the teacher's classroom practices in the form of narrative vignettes of classroom episodes. The final part of the story explores whether the beliefs underlying the teacher's classroom practices match those espoused by the teacher in interviews. In presenting the case stories, I have attempted to allow the teachers and the classroom descriptions to 'speak for themselves' as far as possible. This aims to provide readers with a direct view of what transpired in the TAS classrooms and to allow them to listen directly to the voices of the teachers and students as they were experiencing the TAS.

Looking across all the cases, **Chapter 12** reveals that there were considerable variations in the approaches used by these teachers to fulfil their dual roles as assessors and teachers. Some teachers viewed the scheme primarily as an assessment reform replacing the traditional practical examination. Others viewed it also as a pedagogical reform advocating the teaching of practical work through an investigative approach. A few of them also saw it as a stimulus for their own professional growth. Embedded in these three different interpretations are different conceptions of the notion of *fairness* and different senses of *professionalism*.

These two emerging features of the analysis are discussed in more depth in Chapters 13 and 14. Using a comparative case study approach, in **Chapter 13**, I draw on three cases to probe further into the teachers' different conceptions of fairness, and in **Chapter 14** I use the lens of teacher professionalism to draw on five cases to investigate the reasons underlying teachers' different interpretations of the TAS regulations, and hence their different enactment of the reform.

Chapter 15 first recapitulates the major findings arrived at in each of the earlier chapters, viz. (1) the teachers' varying teaching practices, (2) their perceptions of the TAS, (3) their different views of fairness, (4) their beliefs about science, teaching and learning, and (5) their professional consciousness. Marton and Booth's (1997) theory of the structure of awareness is then used to synthesize the inter-relationships among the findings and to draw some overall conclusions. Implications of the findings for initial teacher education, teacher professional development and educational reform are then discussed. Lastly, some recommendations for future research are provided.

Appendices A and B are academically oriented and written for readers who are more research-oriented. **Appendix A** provides a review of the literature that has shaped and guided the research reported in this book. **Appendix B** gives an account of the research methodology and the process of devising an analytic framework to make sense of the large amount of data collected (which entails, among other data, transcripts of 52 classroom observations and 92 interviews). It culminates with a reflection on some of the methodological issues related to the study.

4. HOW TO USE THE BOOK

None of the cases reported in this book should be regarded as typical or representative in their entirety. Rather, in each case, there are episodes, experiences and emotions with which many teachers will readily identify. As a consequence, instructors of teacher education programs or teacher professional development courses can use the book as a basis for either individual or collaborative teacher development activities. The chapters have been ordered for convenience of presentation. In general, Chapters 2 and 3 should be read first since they provide readers with background information that will enable them to better interpret the case stories. The case stories from Chapters 4 to 11 may be read in any order. Chapters 12 to 15 should be read in that order only after the cases in Chapters 4 to 11 have been thoroughly discussed and reflected on.

Below are some specific suggestions for alternative 'paths' of reading the book, discussing and/or reflecting on the cases. For readers with an interest in exploring the following issues/topics:

- fairness in school-based assessment – Chapters 4, 5, 11 and 13.
- teacher professionalism in education reforms – Chapters 4, 6, 7, 8, 10 and 14.
- the role of practical work in science teaching – the sections on "the role of practical work", "aspects of the nature of science emphasized", "characteristic features of practical work teaching" and "beliefs underlying classroom practices" of the case reports in Chapters 4 to 11.
- the role of school-based assessment in science teaching and learning – the sections on "characteristic features of assessment practices" and "beliefs underlying assessment practices" of the case reports in Chapters 4 to 11.
- the role of metaphors in guiding teachers' practices and its use in examining teachers' beliefs – Chapters 4, 6, 8, 10 and 12.

Whatever your interest is, I invite you to use the cases as thinking devices to uncover and clarify your own educational beliefs and philosophies of education. For some of you, it could be an awakening into consciousness where your familiar daily routines of professional practice suddenly become discordant symbols of conflicts that existed between articulated and unarticulated levels of knowing. To achieve this goal, you are encouraged to consider the following questions, and the like, wherever appropriate:

- How does this case reflect experiences you have had?
- If you were in the teacher's place, what would you do?
- What are the major issues presented in the case?
- What lessons can be learned from the case?
- Identify what you believe should be changed in order to improve the situation? How and why?

In sum, there are just a few voices: teachers', students' and my own, in this book. They speak to a few, but important, issues. My task has been to use the cases and my interpretations of them as a way of opening up the dilemmas teachers face in their classrooms in the context of an assessment reform. I hope that you find the cases as thought provoking as I have found them.

Chapter 2

THE ASSESSMENT REFORM

This chapter provides more information about the context of the study – the Hong Kong Advanced Level (HKAL) Biology TAS – including its development history, its assessment requirements and regulations and some prior studies on it. It is hoped that these highlights of the TAS will provide readers with a background against which they can better appreciate the classroom actions of the teachers reported in the case stories.

1. THE HONG KONG EDUCATIONAL SYSTEM

As described in Chapter 1, the Hong Kong educational system may be characterised as an examination-led system, where what transpires in the classroom is largely dictated by what happens in the public examination halls. Such a criticism can be traced back to 1982 when an OECD review panel was appointed by the then Hong Kong Government to review its educational system. In the panel's report, Hong Kong schools were described as having "obsessive concern" with testing (OECD, 1982). In the course of his/her school career, a child could go through as many as eight sets of selection examinations – from interviews for gaining admission to prestigious kindergartens to the Advanced Level Examination at the end of Secondary 7 (Year 13) for gaining a place at a tertiary institution. In other words, each of these examination hurdles would have an impact on the opening and closing off of certain options for the child at his or her next level of schooling. As a consequence, examinations determine the quality of the educational experiences of teachers and students. Biggs (1996) found that at all stages, "the curriculum, teaching methods, and student study methods, are focused on the next major assessment hurdles" (p.5).

B.H.W. Yung, (ed.), Assessment Reform in Science, 11-18.
© 2006 *Springer. Printed in the Netherlands.*

Since 1982, the number of public examinations in Hong Kong has been dramatically reduced. There remain only two public examinations, one at the end of Secondary 5 and the other Secondary 7 (the Hong Kong Certificate of Education Examination and the Hong Kong Advanced Level Examination respectively). However, the obsession with testing and examination for selection continues. As the then Secretary of the Hong Kong Examinations Authority (HKEA) observes, "In fact, student sometimes stop their teachers from teaching certain topics or materials which are not in the [examination] syllabus" (Choi, 1999, p.412). This echoes the sentiments expressed in the experiential accounts written by both Hong Kong teachers and students of their lived experiences of a particular moment of an examination experience (Pong and Chow, 2002). Expressions like "I breathe deeply"; "trying my best to keep calm"; "my heart sinks"; "panic come over me"; "I try to hold back my tears"; "tears pour down my face"; "I am so nervous about it"; "I heave a sigh of relieve"; and the like, are pervasive throughout the accounts. In sum, there is little doubt that an examination-oriented culture is firmly embedded in Hong Kong, and that examinations are stressful both for the students and the teachers, who know that there is much at stake. In such circumstances, there is a likelihood that attempts to introduce progressive practices may be stifled, unless there is a corresponding change in such high-stakes examinations. Even the HKEA acknowledges in their official publication the fact that "if it is not examined, it won't be taught" (HKEA, 1994, p.80), and much thought has been given to how the examination process can be used to bring about positive and constructive changes to the system. This leads us to the origin of the HKAL Biology TAS – the research context for the study reported in this book.

2. ORIGIN OF THE TAS

Before its reform, the HKAL Biology Practical Examination had long been criticised for several reasons, most of which were related to the large number of candidates taking the examination. For example, because of the insufficient laboratory accommodation, several comparable papers had to be set every year in order to cater for the large candidature. As the candidates had to sit for different sessions of the practical examination, this created a problem in moderating their marks. Because of the large number of parallel sessions and the limited range of practical work that could be included in a three-hour examination, repetition of questions became inevitable, such that candidates sitting for later sessions might have an advantage. Clearly, the problem of moderating marks and the limited scope of practical work that could be tested in the practical examination affected both the reliability and validity of the examination. In addition, because of the stereotyped nature of the examination, candidates tended to concentrate only on work which was likely to be examined. This was reflected in the over-practising of rat dissection on the one hand (it was not uncommon for a student to dissect 20 rats in preparing for the examination), and neglect of ecology fieldwork on the other. These factors had deleterious effects on the curriculum. Thus, there were repeated calls for a better alternative to the practical examination from as early as 1978, and school-based assessment, in one form or another, was among the proposed alternatives.

Two feasibility studies (Edie, 1978; Lau Chang, 1980) on school-based assessment of practical work of students in the HKAL Biology Examination were conducted in 1978 and 1980. However, neither study succeeded in bringing about changes in public examination of practical work in biology. Indeed, attempts at reforming teaching, learning and assessment in Hong Kong have revealed that assessment has usually been the feature most resistant to reform (Morris et al., 2000). This has particularly been the case when attempts to introduce school-based assessment have challenged a traditional emphasis on fairness and objectivity as the main features of the assessment process (Biggs, 1996).

Based on the experiences gained from the earlier attempts, as well as the development of similar schemes operating in overseas countries, a new scheme was designed and a third feasibility study (Yung, 1992) was conducted between 1988-90 by the present author, who was then Subject Officer (1986-93) of the HKEA. The study spanned two years and involved 353 students in 21 schools, who took part in an experimental try-out of the TAS. The students then sat for the 1990 HKAL Biology Practical Examination as normal, and their practical examination scores were compared with those obtained in the TAS. Similar to the two previous studies, the correlation between the two sets of marks obtained by means of the two different modes of assessment (i.e. TAS versus practical examination) turned out to be very low (0.24) and did not support replacing the HKAL Biology Examination with the TAS. However, other data provided by the study suggested that there was a general consensus among the teachers that TAS was the more desirable mode of assessment, in terms of its educational values, when compared with the then practical examination. In particular, the teachers thought that implementation of the TAS would:

- alleviate the problem of the over-practising of rat dissections
- reduce students' examination pressure (with regard to practical work)
- allow a valid assessment of students' practical abilities
- enhance teachers' professionalism and widen their experience

Based on the favourable responses from teachers involved in the trial study, the HKEA decided to replace the practical examination by the TAS starting from the 1995 examination.

3. THE PRESENT SCHEME

Since its inception in 1993 (leading to the first examination in 1995), there have been a number of changes made to the TAS regulations and its assessment requirements. Most notable, is the increasing emphasis on using the investigative approach in teaching practical work. This section aims to highlight some important features of the TAS. As such, a major portion of this section is based on the TAS Handbook for the 1999 HKAL Biology Examination (HKEA, 1997) – the examination for which students participating in this study were prepared.

3.1 General features

The TAS is a mandatory scheme for all school candidates. It aims to enhance the practical approach and to assess practical skills and affective characteristics which cannot easily be assessed by the traditional public examination. It stresses that "the assessment should form part of the normal teaching programme so that the candidates will feel free from the pressure associated with the formal practical examinations" (HKEA, 1997, p.1). To join the TAS, schools have to meet a number of requirements, including the following:

- The student:teacher ratio should not exceed 30:1 at any time.
- Schools should have adequate equipment and laboratory facilities.
- Schools should allow the TAS Coordinator or TAS Supervisor to review the laboratory notebooks and other assignments of students upon request.

TAS Coordinators, who are experienced teachers appointed by the HKEA, monitor 15 schools each. The responsibility of the TAS Coordinator is to ensure that teachers under his/her charge are complying with the requirements of the TAS. The Coordinator will be asked to comment on the marking standards of teachers and the standards of students' TAS work amongst the schools under his/her charge. To do this, the Coordinator has to inspect the teachers' records of work, including lab manuals, assessment checklists, the number and nature of practical work assessed and students' reports, etc. The TAS Supervisor is equivalent to the Chief Examiner of the theory paper, and is responsible for overseeing the implementation of the scheme.

3.2 The assessment areas

In the TAS, teachers must carry out the required *minimum* number of assessments of practical work for each of the specified assessment areas, as summarised in Table 2-1. Of the six assessment areas, the current study focused on two areas, namely (A1) ability to organize and perform practical work, and (B) reporting of investigative work. All the reported classroom observations were conducted in classes where the teacher was carrying out assessment in the above two areas.

Two things are worthy of mention here. First, throughout the two-year A-level course, there must be at least two assessments in which students are required to solve a problem posed by the teacher. These are designated as Type I experiments, in which teachers are expected to provide no further guidance on the design and planning of the investigation other than posing the problem and providing a list of apparatus and materials available for solving the problem in the lab manual. This is in line with the intention of the TAS to promote the investigative approach in teaching practical work.

Table 2-1. Summary of assessment requirements of the HKAL Biology TAS

Assessment areas	Minimum number of assessment	Mark calculation method	Maximum mark
A1 Ability to organize and perform practical work	6	Average of the best 6, of which no more than 2 assessments can be on each of the following: use of microscope and rat dissection	10
A2 Ability to make good-quality biological drawings	6	Average of the best from each type of drawing (whole specimen, L.P. & HP drawings)	5
B Reporting of investigative work	6	Average of the following 6 reports: the best 2 Type I reports and the best 4 from the rest of Type I and all Type II reports	30
C Experience in rat dissection	Experience only		3
D Experience in ecology field work	Experience only		3
E Attitude towards study of biology	Assessed throughout the A-level course		5

Indeed, teachers are encouraged "to allow students to investigate their own problems instead of posing a problem for them" (HKEA, 1997, p.25). Second, in order to tackle the problem of the so-called 'right-answer syndrome' among the students when they carry out practical work, and to inculcate in them an appropriate scientific attitude, the scheme specifies that "Students should be made aware that when carrying out an investigation, it sometimes happens that they may obtain results that deviate from their expectation or contradict with the theory. In such a case, they should be encouraged to present the anomalous results in an honest way, and consider whether these results suggest the need for modifying the experimental procedure, or whether they may provide evidence for rejecting the original hypothesis. Thus a mark penalty should not be made on assessment area B solely for such results..." (HKEA, 1997, p.11).

3.3 Making assessment

In line with the idea of continuous school-based assessment, it is recommended that the TAS assessments should be spread over the two years of the A-level course. Teachers are encouraged to carry out the assessment as part of the normal teaching programme, but they are not barred from assessing certain exercises in a more formal way, for example, in the form of a practical test. Teachers are required to inform their students at the beginning of the A-level course that parts of their practical assignments during the course will be used for assessment purposes. However, whether teachers inform students that their work is being assessed for examination purposes on a specific occasion is left to the discretion of individual teachers. Teachers are required to record their grading for individual assessments on the students' work as part of the normal feedback to the student. This is line with the TAS intention of providing students with continual feedback.

Regarding the actual assessment practices, three approaches are suggested: (1) awarding marks by following a checklist, (2) marking by impression, (3) supplementing (1) and (2) by oral interaction while observing the students work. Provision is also made to allow teachers to offer help or advice to students who experience problems with any work being done for assessment. An appropriate amount of marks *may* be deducted in such cases, depending on the amount of help given. The scheme also requires teachers to exercise control and supervision over all work on which assessments are carried out, in order to ensure that the work assessed is the candidate's own work. However, it does not bar teachers from asking students to complete the assessed work outside class time, for example, completing it at home, if the circumstances arise.

3.4 Moderation of TAS marks amongst the schools

The TAS is intended to reflect the final standard of work achieved by a student, so that an early assessment should be replaced by a better one on a later piece of work. It is not, however, intended that the same exercise can be assessed, and then revised and resubmitted by the candidate in order to gain a higher mark. The sum of the best assessments will provide an overall mark for each student, and these marks will provide a rank order of the students in each school. The final TAS marks submitted by a school will be subjected to a moderation procedure which is not intended to alter the teacher's rank order of his/her students within a school, as established by their TAS marks. The marks given by a school will, however, be adjusted when the moderation shows it necessary to equate the standards of marking of the individual school with those of the A-level entry as a whole. Given the small class sizes (10-15 students per class) in A-level Biology, the moderation is basically a combination of statistical moderation and moderation by inspection. This process provides weight for the TAS Coordinators' comments on the teachers' standards of marking.

4. PRIOR STUDIES ON THE HKAL BIOLOGY TAS

In reviewing the special features of the HKAL Biology TAS, the then Chairman of the HKAL Biology Subject Committee, Pang (1992), highlighted some of its educational values. For example,

- TAS can achieve better assessment by allowing more valid and reliable assessment of students' practical skills as well as a wider measurement of students' achievement of educational objectives, in particular, students' affective characteristics.
- TAS can motivate students to improve their own performance continually because only the best marks are taken for each student.
- The teacher may also provide more feedback to students based on their TAS results.
- TAS has a liberating influence on teaching. Since there is no further need for drilling for particular types of practical work required by practical examinations, more varied, meaningful and relevant experiments and projects can be conducted. This will allow for better integration of theory with practical work.

Teachers participating in the trial study of the Biology TAS expressed similar views towards the educational values of the TAS (Yung, 1992). Overall, the findings largely echoed what Pennycuick (1990) identified in analyzing teacher assessment systems in some developing countries, namely, that it:

- enhances validity of assessment,
- allows for integration of curriculum, pedagogy and assessment, and
- serves a broader range of assessment functions, in particular, it emphasises formative functions.

Pang (1992) also argued that implementation of the TAS would provide opportunities for the professional development of teachers via various in-service training activities. In a study conducted two years after the first implementation of the TAS (Yung, 1995), most teachers (N=222, representing 67% of the total population) confirmed that the TAS had contributed significantly to their professional development. On the knowledge side, teachers felt that they had come to realise the importance of assessing both the product and process of students' practical work, the importance of assessing students' affective characteristics, and the importance of integrating teaching with assessment. They thought that they had also gained a better insight into the role of practical work in biology teaching and a better understanding of the functions of both formative and summative assessment. With regard to skills, the teachers reported that the TAS had contributed significantly to their skills in assessment and to their ability to inculcate in students the appropriate affective qualities. In addition, teachers felt that they had gained improved access to useful ideas for teaching and assessment. Overall, the teachers acknowledged the contribution of the TAS to their professional development and thought that they had become more accountable to their students and professional colleagues. However, it should be noted that the studies cited above (Yung, 1992, 1995) were quantitative studies where teachers were asked to express their views on *a priori* statements using a Likert scale. Such a reductionist approach to the problem did not reveal much about what was actually taking place inside the classrooms. Indeed, the naturalistic qualitative study reported in this book uncovers quite different stories.

Chapter 3

TEACHERS' ENACTMENT OF THE REFORM

This chapter analyses the teachers' classroom practices so as to generate a combined account of the practices of all the ten teachers in a way that can capture the interrelated aspects that made up their practices. This combined account provides a database of the different aspects of the teachers' practice, which can be drawn on as evidence for subsequent analysis. Subsequent chapters look at the classroom practices of individual teachers in a more focused manner, based on the information presented in this chapter.

1. PROBLEMS ENCOUNTERED IN DRAWING UP A COMBINED ACCOUNT OF THE TEACHERS' PRACTICES

The number of practical work sessions observed for each teacher ranged from four to eight, and the sessions varied considerably in terms of the following variables:

- Whether it was an assessment practical or not
- The nature of the practical work itself
- The duration of the practical session which ranged from 105 – 210 minutes
- The amount of dialogue between the teachers and their students which ranged from 87 – 788 dialogical text units* (DTUs).

*(*Explanatory Note: A DTU refers to the part of a dialogue exchange [between the teacher and the student(s)] which is followed or separated by a natural break or pause in speech. That means, the number of dialogical text units gives a rough indication of the amount of interaction between a teacher and his/her students during the practical.)*

B.H.W. Yung, (ed.), Assessment Reform in Science, 19-38.
© 2006 *Springer. Printed in the Netherlands.*

All the above variables (amongst others) may have a bearing on the foci and pattern of teacher-student (T-S) interaction, and must be dealt with properly when drawing on data to produce a combined account that allows a fair comparison of the teachers concerned. Since the present study was a naturalistic study, there were only certain ways in which I, as the researcher, could manipulate any of the above variables. The first two variables were accounted for through sampling the practical sessions to be included in the comparative analysis. Only assessment practicals were considered in the comparative analysis of the classroom data. Three assessment practicals from each teacher were sampled, such that the nature of the sampled practicals could be as similar as possible for each of the teachers.

The third and fourth variable cited above [i.e. the varying lengths (105 – 210 minutes) of the practicals and the varying amounts of T-S interaction (87 – 788 DTUs) taking place] were dealt with by expressing the foci and types of T-S interactions as *percentages of the total DTUs* recorded in each of the lessons for a particular teacher, instead of the *actual* number of DTUs recorded for each of the categories in the coding scheme. To further enhance the comparability of the data obtained from different teachers' lessons, the analysis of the pattern of T-S interaction was restricted to only those dialogues that were directly related to the teaching of practical work *per se*. That is, other dialogues related to the normal classroom routines or otherwise were not included in the analysis.

Prior to the study, there was no *a priori* coding system for the classroom data (i.e. the T-S interactions). Instead, a system was formulated as a result of several attempts at data examination, reduction and display (see Appendix B for a detailed description). The coding system so derived comprised two dimensions: (1) ***Foci*** of T-S interaction, and (2) ***Types*** of T-S interactions. (The relative occurrence of the different types of T-S interaction in a teacher's class constitutes the ***pattern*** of T-S interaction for that particular teacher.)

It was evident from the data that there were dual aspects of classroom events taking place in the TAS practicals – teaching and assessment. These reflected the two roles of teachers in the context of the TAS – as teachers and as assessors. For this reason, the dimension of foci of T-S interaction in the coding system was further divided into three subgroups, namely, (1a) T-S interactions focusing on *assessment-related issues*, (1b) T-S interactions focusing on the *teaching of practical work per se*, and (1c) *other foci* of T-S interaction.

Both (1a) and (1b) were further divided into sub-categories detailed in Table 3-1. In brief, the coding system was so constructed as to comprise only those variables of foci and types of T-S interaction which appeared to stand out as related to the teachers' beliefs and different understanding/

Table 3-1. Summary of the coding system for the classroom data

(1)	Foci of T-S interaction		
	(1a)	Focusing on assessment-related issues	
		i.	Same time allowance
		ii.	Intervene in students' discussion
		iii.	Refrain from helping students
		iv.	Concern with marks
	(1b)	Focusing on teaching of practical work *per se*	
		i.	Scientific thinking
		ii.	Biology theory behind the practical work
		iii.	Scientific attitudes
		iv.	Procedure, demo + safety
		v.	Apparatus + materials
		vi.	Progress of work
	(1c)	Other foci of T-S interactions	
(2)	Types of T-S interactions		
	(2a)	T guides S to answer	
	(2b)	T gives answer directly	
	(2c)	T assists S	
	(2d)	T gives direction	
	(2e)	T asks S's progress	

perceptions they had of the TAS. Essentially, each DTU would be assigned two codes: one for the *focus* of T-S interaction and another one for the *type* of T-S interaction.

2. THE COMBINED ACCOUNT OF TEACHERS' PRACTICES IN THE TAS

Given the relatively small number of TAS practicals (i.e. three) sampled from each of the teachers for analysis, the following combined account of teachers' practices in the TAS should be read with caution. Nevertheless, I believe that such a combined account can serve the purpose of giving readers a general idea (in quantitative terms) of certain comparable tendencies among the teachers' practices in the TAS.

2.1 Amount of T-S interaction in the TAS practicals

Figure 3-1 shows the relative amount of interaction (in terms of number of DTUs per lesson) between the teachers and their students during the TAS practicals. The differences among the different classrooms in this respect were quite appreciable. The number of DTUs per lesson ranged from as many as 718 in Alan's classroom to as few as 132 in John's classroom. A general trend, illustrated by Figure 3-1, was that teachers on the right hand side of the chart tended to interact less frequently with their students during the practical than their counterparts on the left hand side of the chart.

Readers may immediately query that the trend reported above is, in itself, rather meaningless unless the relative positions of the teachers on the horizontal axis of the bar chart carry some meaning. Indeed, the positions of the ten teachers on the horizontal axis of the bar chart have not been assigned randomly, but with a purpose in mind, such that this chart can be easily cross-compared with subsequent charts concerning the same teachers (see Figures 3-2 to 3-6). In fact, the original rationale for determining the order of the ten teachers on the horizontal axis of the bar charts was the *decreasing amount of T-S interaction*. Such a rationale was arrived at based on the general feeling obtained through many of the class observations that those teachers who were more concerned with the TAS assessment *per se* were also those who tended to interact less with their students during the assessment practicals. However, as the data emerged, it became obvious that it would be best to position the teachers on the horizontal axis of the bar charts according to their *increasing concern with assessment-related issues* during the practicals. The rationale for using this criterion will be explained in a later section, together with a detailed description of the assessment-related issues involved. This order of the teachers on the horizontal axis in all the bar charts will be kept the same throughout the data reporting.

In general, then, Figure 3-1 indicates that the teachers who were more concerned with assessment-related issues (those situated more on the right-hand side of the chart) tended to interact less often with their students during the practical than their counterparts who were less concerned with assessment-related issues (those situated more on the left-hand side of the chart). Subsequent analysis revealed that the low levels of T-S interaction among the *assessment-oriented* teachers (those who were more concerned with assessment-related issues) were due to their tendency to refrain from offering assistance or discussing with their students during the TAS practicals. This was because they saw this as essential in maintaining fairness of the assessment (see also Chapter 13).

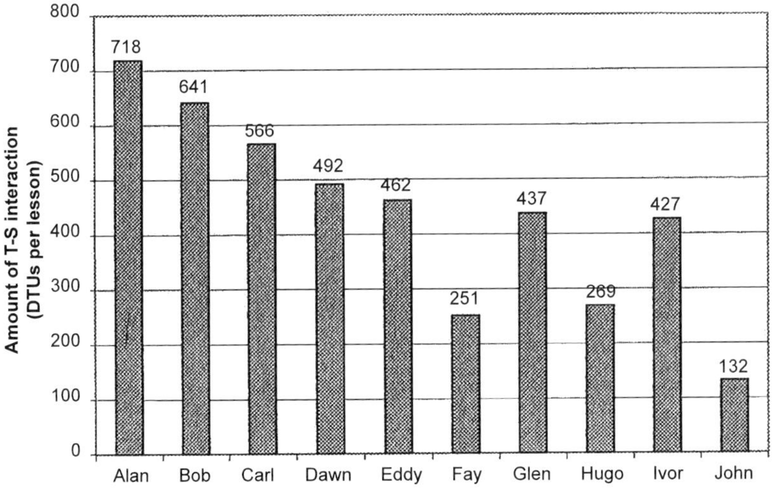

Figure 3-1. Amount of T-S interaction during the TAS practicals

2.2 Different foci of T-S interaction in the TAS practicals

In examining the interactions between the teachers and the students, consideration of the nature of the focus of the discussions is at least as important as the amount of interaction. Figure 3-2 shows the emphases which different teachers put on the different areas, namely, (1a) assessment-related issues, (1b) the teaching of practical work *per se* (PW teaching-related), and (1c) other foci of interactions when interacting with their students during the TAS practicals. Every effort was made to avoid assigning two codes to a particular dialogue regarding its focus of T-S interaction. In no cases, was double coding more than 2% of the total number of DTUs of the lessons recorded.

As would be expected, a large proportion of the T-S interactions (ranging from about 83% to nearly 96%) concerned aspects related to the teaching of practical work *per se* in all the cases. A small proportion of T-S interactions (ranging from about 2% to 9%) were related to 'other foci' like taking attendance, distributing lab manuals, collecting lab reports, social chatting, etc. There was no observable relationship between the amount of T-S

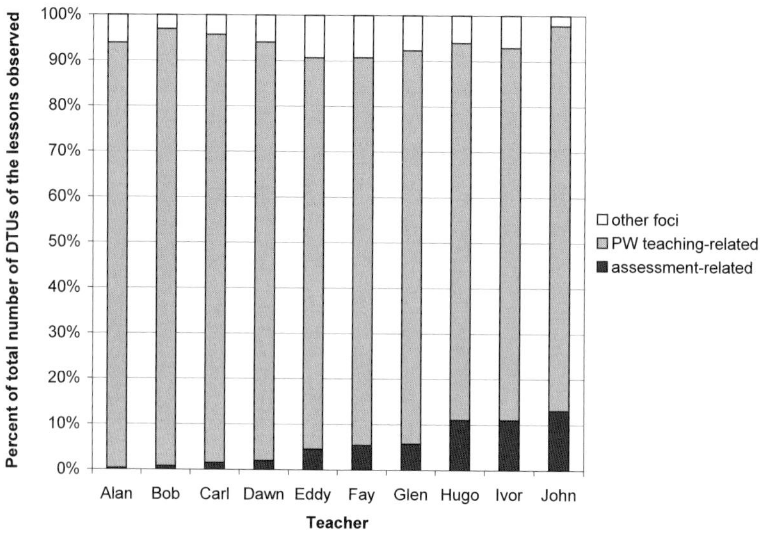

Figure 3-2. Foci of T-S interaction during the TAS practicals

interaction with regards to these 'other foci' and the teachers' concern with assessment-related issues (i.e. their position on the horizontal axis of the chart). On the contrary, the amount of T-S interaction concerning assessment-related issues (ranging from 0.3% to 13% when one moves from the left-hand side towards the right-hand side of the chart) was found to be negatively correlated with the amount of T-S interaction concerning the teaching of practical work *per se*. In other words, the more assessment-oriented the teacher was, the less he/she would interact with his/her students (in terms of % DTUs of the lessons observed) during the teaching of practical work, or vice versa. In the following two sections, the two key foci of interaction (assessment-related issues and practical work teaching-related issues) will be explored in more detail.

2.3 Foci of T-S interaction related to assessment issues

In the course of the field study, it became obvious to me that in some classrooms, students' obsessive concern with assessment marks (sometimes aggravated by the teachers' classroom practice) had a profound influence on both the foci and the variety of types or pattern of T-S interaction. Figure 3-3 shows the different foci of interaction under this category for the ten teachers

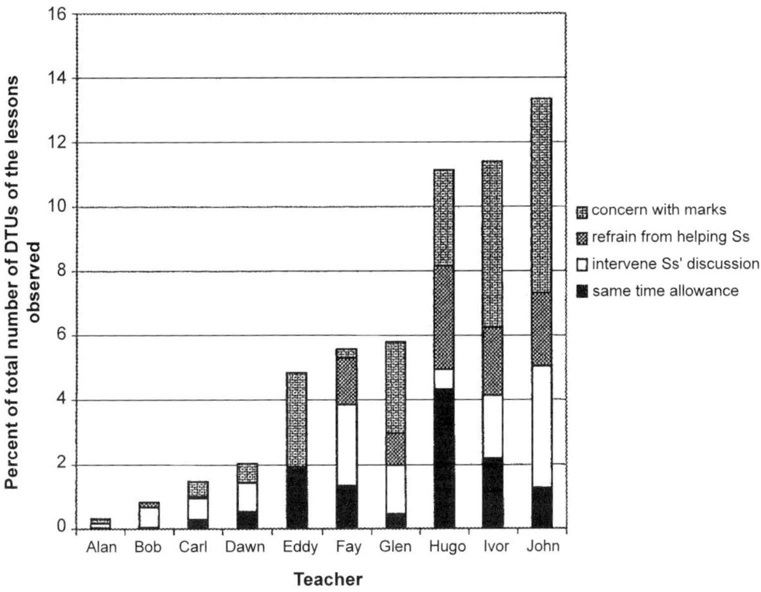

Figure 3-3. Foci of T-S interaction related to assessment issues

in this study in terms of percent of the total number of DTUs of the lessons observed (% total DTUs).

Before studying Figure 3-3 more closely, it is necessary to define the symbols used in the legend of the chart, namely:

- same time allowance
- intervene in students' discussion
- refrain from helping students
- concern with marks

These represent the different subcategories of interaction that were in one way or another related to the students' or the teachers' concern with issues related to the TAS assessment, especially *fairness* of the assessment. In order to let readers develop a more 'qualitative' feel for the situation, each of these sub-categories will be illustrated by a representative classroom episode. In brief, the following episodes serve a dual function – (1) to define the symbols used in the legend of Figure 3-3; and (2) to give readers a feel for the classroom situations in the context of TAS. Table 3-2 gives the transcript conventions used in the classroom episodes.

Table 3-2. Transcript conventions used in the classroom episodes

Convention	Meaning
TWC	Teacher's talk that is directed at the whole class
T	Teacher's talk
S	Student's talk. In cases where there is more than one student involved in the dialogue, the students are designated as S1, S2, S3 and so on.
[ ]	Words enclosed in 'square' parentheses aid understanding and readability of the dialogue.
( )	Words enclosed in parentheses indicate non-verbal actions, e.g. writing on the board
(?)	Uncertainty in the transcription
[...]	Short pause, typically of about 3 seconds
...	Indicates a section of the dialogue from within the excerpt has been omitted

2.3.1 Same time allowance

Eddy, Fay, Hugo, Ivor and John spent 1 to 4% of their T-S interaction in this area, that is, ensuring that all students were given the same time allowance for completion of the assessment task. This group of assessment-oriented teachers thought that this would help to ensure fairness of the assessment. They tended to see the TAS assessment as equivalent to the traditional examination, which is conducted within well-defined time limits. Thus, they would often not start a lab session until all the students had arrived, or until the whole class was ready to start. The following is a representative episode of this kind.

Episode 3-1 (Hugo) – For fairness sake, let's wait for the monitor

After taking roll call, the teacher, Hugo asked the monitor to help him to return the attendance register to the General Office, and then made the following announcement to the whole class:

TWC: For fairness' sake, let's wait for the monitor before I give you the lab sheets.

When the monitor returned after three minutes, the teacher distributed the lab sheets to the class and then made the following announcement:

TWC: I will not give you any reminder of the time left. When the bell rings, I will begin a count down of 60 seconds. That means, after the bell rings, you have to submit your report within one minute. After that, I will not accept any more reports. So, you have to plan your time well.

Among the ten teachers, Hugo was the most stringent in requiring students to observe the deadline for completion of work and handing in the reports. This accounted for the 4% of T-S interaction (which was the highest amongst all the ten teachers) that he had with his students on this aspect. In addition to his concern with maintaining fairness of the assessment, as discussed above, this was also related to his using this as a 'tool' for training students to prepare for their future working life – that is, to "learn to meet deadlines" (see Chapter 9 for details).

2.3.2 Intervene in students' discussion

Fay, Glen, Ivor and John spent 1 to 3 % of their T-S interaction in this area, that is, intervening to stop students from discussing among themselves during the TAS practicals. This group of assessment-oriented teachers thought that preventing discussion would ensure fairness of the assessment. They tended to see the TAS assessment as equivalent to the traditional examination, during which students are not allowed to talk to each other, otherwise, it would be regarded as cheating, as revealed in the following representative episode.

Episode 3-2 (Hugo) – I could take this as cheating

The teacher, Hugo, noticed that two students were discussing with each other. He intervened and talked to them in a very serious manner:

T: You two. No discussion. Do you know that I could take this as cheating? You know. When you sit for an external examination, you can't discuss with your neighbour… So, Clarie, next time, when your neighbour asks you a question, you'd better refuse to answer her question. This is for your own benefit.

The very serious manner in which the teacher, Hugo, handled the above situation is not revealed in the quantitative figures presented in Figure 3-3. This is because he spent less than 1% of his interaction intervening in students' discussion, the reason perhaps being that he took a serious stance on this issue, and students knew about it. Hence, they tended to refrain from discussing with each other. Thus, the low percentage of Hugo's T-S interaction in this area, relative to that of other assessment-oriented teachers, might just be an indication of the low frequency of occurrence of student discussion generally in his class (which was in fact true). Nevertheless, this episode does illustrate how quantitative figures may sometimes fail to reveal important information. More detailed and qualitative descriptions of the teachers' classroom actions can be found in their individual case stories in Chapters 4 to 11.

2.3.3 Refrain from helping students

Fay, Glen, Hugo, Ivor and John spent 1 to 3 % of their T-S interaction in this area, that is, during their interaction with students, they expressed an unwillingness to help in solving the problems raised by their students during the TAS practicals. Once again, this group of assessment-oriented teachers thought that this was essential in order for them to maintain fairness of the assessment. They tended to equate the situation with that of the traditional examination, where the examiner would not give any help to the candidates on the spot. Such a phenomenon was particularly obvious in Hugo's case, in that he often refrained from offering to help students, sometimes despite students' repeated requests. Below is one of the representative episodes.

Episode 3-3 (Hugo) – I don't think it is appropriate for me to tell you

It was a practical on paper chromatography. Hugo was circulating around the lab. One student approached him and asked him a question about the correct way to measure the distances travelled by different colour spots on the chromatogram:

S: When I measure the distance travelled by the different colour spots, shall I measure the centre part or at the top part [of the colour spots]?

T: It is something that is to be included in your report. Therefore I don't think it is appropriate for me to tell you.

2.3.4 Concern with marks

The assessment aspects were dominant for all students and teachers in this study. Thus, it is not surprising that all teachers did touch upon the issue of assessment marks in one way or another during the TAS practicals. Their T-S interaction relating to this aspect ranged from less than 1% in Alan's case to more than 6% in John's case. Negotiations and/or disputes between the teacher and individual students concerning TAS marks awarded for a certain piece of work were more common in the TAS practicals of the more assessment-oriented teachers (e.g. Eddy, Glen, Hugo, Ivor and John). The following is a representative episode.

Episode 3-4 (Eddy) – Are you going to deduct marks from me?

Eddy was pointing out to a student on the spot that he had taken more than the specified amount of materials and that marks would be deducted from him:

T: You too [getting more than the specified amount of solution]. What's your class number? Next time, remember that 10 ml is just that much! 10 ml is just that much! You have taken four times more than enough.

S: Are you going to deduct marks from me?

T: Of course, you have had too much material.

2.3.5 A summary on T-S interaction related to assessment issues

As indicated in Figure 3-3, of the ten teachers, Alan (and his students) was the least concerned with the assessment-related issues while John was the most concerned. Altogether, three teachers (Hugo, Ivor and John) spent up to 10% or more of their interactions with students on issues related to this aspect. Ten percent or more was definitely not a low figure when compared with Alan and Bob, who spent less than 1% of T-S interaction on the same area. In subsequent analyses, readers will find out that these three teachers carried out the TAS assessments in a manner that was very similar to the procedure in a traditional public examination, where the implementation of fair procedures would be a prime concern. This explained the relatively high proportion of T-S interaction in this area for this group of assessment-oriented teachers. This was very much related to their perception of the TAS as primarily an assessment reform replacing the traditional practical examination (see individual case stories in Chapters 4 to 11 for more details).

2.4 Foci of T-S interaction related to teaching of practical work *per se*

Figure 3-4 shows the different foci that each teacher emphasized [in terms of % total DTUs (PW)] in their teaching of practical work *per se*. [DTUs (PW) refer to those DTUs that are specific to the teaching of practical work *per se.*] Before studying Figure 3-4 more closely, it is important to define the symbols used in the legend of the chart and to provide readers with a more 'qualitative' feel for the situation. Representative episodes will be provided in the following six sections to illustrate the meanings of the different *foci* of T-S interactions that are related to the teaching of practical work *per se*, namely:

- *Scientific thinking* – aspects dealing with concepts underlying scientific research
- *Bio theory behind PW* – aspects related to biology theories underlying the practical work
- *Scientific attitude* – aspects related to inculcating in students appropriate scientific attitudes
- *Procedure, demo + safety* – aspects related to the procedural and operational side of the practical work; the focus of the interaction is on the question of 'how' and does not touch upon the underlying biology theory or concepts underlying scientific research. That is, the related question of 'why' is not addressed.
- *Apparatus + material* – aspects related to provision of apparatus and materials for the practical work
- *Progress of work* – focusing on students' progress of work but without ensuing discussion

Wherever appropriate, the same episodes will also be used to illustrate the different *types* of interaction (see Figure 3-5), namely:

- *T guides S to answer* – in response to a student's question (or one raised by the teacher himself/herself), the teacher guides the student to arrive at the answer through discussing with the student
- *T gives answer directly* – in response to a student's question (or one raised by the teacher himself/herself), the teacher provides the answer to the student directly

- *T assists S* – the teacher offers assistance to a student without any request from the student
- *T gives direction* – the teacher gives explicit instructions to a student on how to carry out a certain task
- *T asks S's progress* – the teacher inquires about a student's progress of work without ensuing discussion

2.4.1 Scientific thinking (Concepts underlying scientific research)

As evident from Figure 3-4, teachers on the left hand side of the chart tended to place more emphasis than those teachers on the right hand side on discussing with students concepts underlying scientific research during the TAS practicals. As mentioned before, teachers on the right hand side of the chart were those who were more assessment-oriented (see Figure 3-3). These teachers saw the TAS primarily as an assessment reform replacing the traditional practical examination. This is quite different from those teachers on the left hand side of the chart, who saw the TAS as both an assessment reform and a pedagogical reform emphasizing the investigative approach. These teachers spent a larger proportion of their T-S interaction on developing students' scientific thinking and concepts underlying scientific research. Incidentally, the three teachers (Alan, Bob and Carl) who put more

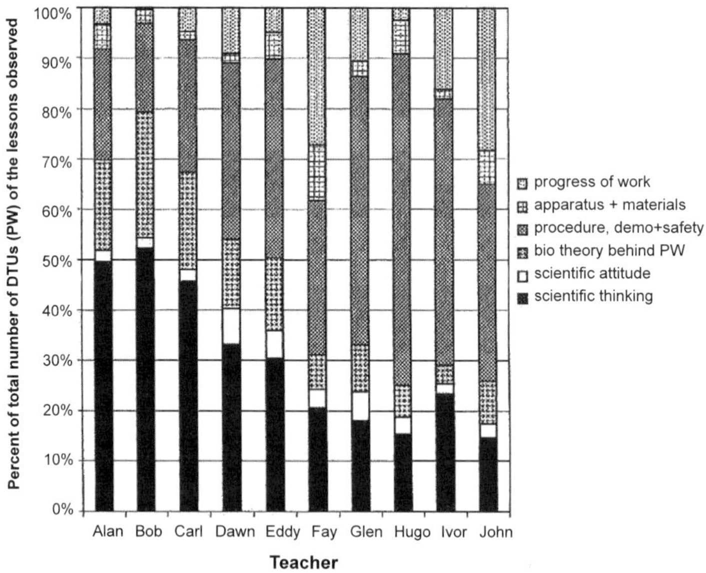

Figure 3-4. Foci of T-S interaction related to teaching of practical work *per se*

emphasis on this aspect of their practical work instruction [in terms of % DTUs (PW)] were those who also had a great interest in the philosophy of science. These teachers also spoke of the role of practical work in school science education in a very articulated manner (see Chapters 4 to 6).

Below is a representative episode where the focus of interaction was centred on developing in students a scientific way of thinking and a mastery of the concepts underlying scientific research. In this case, it was the relationship between the concept of accuracy and the experimental design.

Episode 3-5 (Alan) – Think more carefully

It was a practical on the determination of water potential of potato tissue. A student found the lengths of the potato strips too long to be completely immersed by the sucrose solution in the petri dishes. He raised the problem with the teacher, Alan:

S: Sir, the strips cannot be completely immersed in the solution.

The teacher referred the student to the relevant part of the lab sheet and said:

T: [Here it says:] Prepare potato strips of suitable length. Suitable length! What is suitable length? You have to judge what is the suitable length.

T: In fact, what is the parameter that you are going to measure?

S: Mass.

T: Okay. Suppose, we use potato strips of a shorter length, what will happen to its percentage change?

S: It will become larger.

T: Is it? Think more carefully. Think more carefully.

T: It will be the same. The percentage change will remain the same.

S: Is it that I have to use a longer length?

T: Why then do you have to use a longer strip?

S: The error will become smaller.

T: Why will the error become smaller?

S: [...]

The *type* of interaction represented in this episode is coded as '*T guides S*' i.e. the teacher is guiding the student to the answer through discussing it with him.

2.4.2 Scientific attitude

As revealed in Figure 3-4, there is no discernable pattern among the ten teachers regarding the relative proportion of T-S interaction [in terms of % DTUs (PW)] that was aimed at developing appropriate scientific attitudes in students. In most cases, the teachers required their students to report faithfully on the results they obtained during an experiment, and then try to seek explanations for the 'unexpected' results. The following is one of the representative episodes.

Episode 3-6 (Alan) – Explain your expected results and unexpected results?

S: Sir, it seems that this is not the normal result.

T: You mean this is not your expected result?

S: Yes.

T: But anyway, you got a set of results, right? The fact is that you've got some results, be they accurate or not, okay? You try to record the results first and then try to give some kind of explanation for them.

S: Explain what?

T: Explain your expected results and unexpected results. You said that it is unexpected. This is not what you expect. So, try to explain why. You should have explanations for that.

In the above case, the teacher was providing the student with an explicit instruction, that is, to report the results faithfully in his report. Thus, it is classified as belonging to the category of '*T gives direction*' regarding the *type* of interaction in the teaching of practical work.

2.4.3 Biological theory behind the practical work

As revealed in Figure 3-4, teachers who were less assessment-oriented tended to spend a larger proportion of their T-S interaction [in terms of % DTUs (PW)] on teaching the biological theory underlying the practical work than their counterparts who were more assessment-oriented. The latter group of teachers avoided interacting with students in this respect because they feared that this might jeopardize the fairness of the assessment. This was because they thought that discussing the biology theory underlying the practical work with students would give away part of the answers they expected to see in students' lab reports. On the other hand, teachers who were less assessment-oriented tended to feel less constrained in this respect, as shown in the following episode. In the following case, the *type* of interaction is categorized as '*T guides S*', where the teacher did not answer the student's question directly but was just guiding him through it.

Episode 3-7 (Alan) – Are these strips of unequal water potential?
It was a practical on investigating the water potential of potato tissue. Students were provided with some potato strips for the experiment. A student asked the teacher the following question when collecting the materials back to his workstation:

S: Are these strips of unequal water potential?

T: Ah! For this question, you have to comment on it. Whether they are of equal water potential or not? Do you think you can cut 180 potato strips from a single potato?

T: This is a good question – Are they of equal water potential?

T: I don't know. The fact is that a single potato cannot give you 180 potato strips.

2.4.4 Procedure, demo + safety

T-S interactions that fall under this category differ from the previous categories in that they are concerned with the 'operational' and 'procedural' aspect of the practical work. In other words, they are concerned purely with the aspects of *how* and *what*. The focus of the interactions does not touch upon the theory part of the activity or questions of *why*. Examples of this kind of interaction included: introduction of the problem to be investigated, reminders on laboratory safety issues, demonstrations on the use of instruments and the technical know-how of certain experimental procedures, etc. The following episode is an example of this category. The *type* of interaction is classified as '*T gives direction*', where the teacher gave explicit instruction to the students on what to do or how to carry out a certain task.

Episode 3-8 (Ivor) – Button up your lab gown
It was a practical on the estimation of reducing sugar content in a piece of grape. The following were instructions given by the teacher, Ivor, to his students at different times of the practical on the laboratory procedures and safety measures, etc.:

T: Didn't I tell you in the pre-lab briefing that you should keep the bench surface tidy and clean? You see there are a lot of things lying around.

T: It's rather dangerous that you put the test tubes too close to the edge of the bench. Put them in a safe place.

T: Roll up your sleeves. They are touching the beakers, etc. Beware of knocking them over.

T: The water bath is going to topple over soon. How come you would allow this to happen?

T: Button up your lab gown.

As revealed in Figure 3-4, the general trend observed among the ten teachers was that the more assessment-oriented teachers tended to spend a larger proportion of their T-S interaction [in terms of % DTUs (PW)] in this area during the TAS practicals than their counterparts.

2.4.5 Apparatus + materials

The interaction here is relatively straightforward. Dialogues of this category are concerned with the arrangements for provision of apparatus and materials for the practical work. There was no appreciable trend observed among the ten teachers. The variations among the teachers could be due to a number of factors, including the amount of apparatus and materials used in a particular practical, or whether students were assigned their own set of apparatus throughout the whole term or not. The following is a representative dialogue of this category, and the *type* of interaction is classified as '*T gives direction*'.

Episode 3-9 (Bob) – Here are most of the apparatus that you'll need

TWC: Here are most of the apparatus that you'll need. You may need some more glassware from your own tray.

TWC: Graph paper is over here. You may need to make some graphs. Stands are over there under the bench. You have extra clamps here.

2.4.6 Progress of work

No appreciable trend was observed among the ten teachers with respect to inquiring about students' progress of work. Nevertheless, the assessment-oriented teachers tended to spend a relatively larger proportion of their T-S interaction on this [in terms of % DTUs (PW)]. This is because T-S interactions that fall under this category can easily be developed further if the teacher so wishes. For example, after asking about the student's progress of work, the teacher can easily follow up with further questions related to the underlying theory. If that is the case, this particular segment of T-S interaction would then be classified as 'bio theory behind PW' rather than 'progress of work'. Thus, the relatively higher proportion of T-S interaction in this area exhibited by the assessment-oriented teachers, who did not tend to develop questions, might be explained by their inclination to discuss less with the students so as to avoid giving them hints about the assessment task. This was just the opposite to the case of Bob, who often initiated discussions with individual students, on the procedure or interpretation of the experimental results.

In fact, there were different intentions behind the teachers' enquiring about the students' progress of work. For example, some teachers were clearly collecting information for assessment purposes, as revealed in the following episode.

Episode 3-10 (Hugo) – Oh, yes, before that, I want to check it

T: Have you obtained the results?
S: Yes
T: Oh, yes, before that, I want to check it. Em. Tell me what is the concentration [of the solution] for the different tubes.
S: This is 20 percent …
T: So, this is the lowest concentration and the highest here? Okay.
The teacher carried on inspecting the tubes while interacting with the student:
T: And what is your data? Did you write down your data?
S: Yes.
T: Can I have a look at the dilution table?
S: Okay.
T: So, what's that? Representing the tube number?
S: Yes, tube number.
T: Okay. And this represents?
S: The amount of water I have to add so as to dilute it to the required concentration.
T: Good. Thank you.

In other cases, however, it was rather difficult to ascertain the intention behind the teacher's action, as illustrated by the following episode.

Episode 3-11 (Fay) – How is it going?

T: How is it going?
S: The liquid drop has moved a lot [along the capillary tube of the respirometer].
T: What's the movement?
S: From here to here.
T: And for how long?
S: 15 minutes.
T: How about in the other setup?
S: Just 0.2 cm.

In this situation, it was most likely that the teacher was just 'doing something', with no specific purpose in mind, or was engaged in general monitoring, with no assessment-oriented motive. As Carl put it, "It is just natural enough and is sort of moral support to the students."

2.4.7 A summary on the foci of T-S interaction related to teaching of practical work

Figure 3-4 shows the different foci that each teacher emphasized [in terms of % DTUs (PW)] in their teaching of practical work *per se*. Four general trends were observed. First, teachers who saw the TAS as both an assessment reform and a pedagogical reform tended to emphasize discussion of concepts underlying scientific research during the TAS practicals more than those teachers who saw the TAS primarily as an assessment reform.

Second, the more assessment-oriented teachers also tended to spend less of their T-S interaction on the biological theory underlying the practical work than their counterparts. This was because they thought that in doing so, they would give away part of the answers which they would expect to see in students' lab reports, and hence jeopardize fairness of the assessment. Third, for a similar reason, the assessment-oriented teachers tended to spend a relatively larger proportion of their T-S interactions in *just* enquiring about students' progress of work, but did not develop these discussions further to cover the relevant theories and concepts underlying the practical work. These teachers were inclined not to discuss the underlying theories with the students because they did not want to give away answers to the assessment task, for fear of affecting the fairness of the assessment. Fourth, as a corollary of putting less emphasis on teaching related to concepts underlying scientific research and the biology theories underlying the practical work concerned, the assessment-oriented teachers tended to interact with their students more on the procedural and operational sides of the practical compared with their counterparts who emphasized the former aspects more.

2.5 Patterns of T-S interaction related to teaching of practical work *per se*

Since it is rather tedious to code for all dialogues on the *types* of interactions (cf. *foci* of interaction), I decided not to go into the analysis of the *pattern* of T-S interaction for all foci of interaction but only for those related to the teaching of practical work *per se*. The latter analysis is essential if we want to explore the relationship between a teacher's pattern of T-S interaction in the teaching of practical, and his/her beliefs about the nature of science, about the role of practical work, and about teaching and learning – which is one of the purposes of the current study.

Figure 3-5 shows the *patterns* of interaction [in terms of % DTUs (PW)] of the ten teachers in their teaching of practical work *per se*. As mentioned previously, the focus of the present analysis is just on the teaching of practical work, only those dialogues with an element of *teaching* (and *learning*) through the medium of practical are included. Those related to assessment issues are not included in the analysis.

Two trends regarding the pattern of interaction in the teaching of the practical amongst the ten teachers can be noted in Figure 3-5. Firstly, teachers who saw the TAS also as a pedagogical reform (those on the left hand side of the chart) tended to guide students to think more when interacting with them than those who viewed the TAS primarily as an assessment reform (those on the right hand side of the chart). As a corollary, these teachers also tended to give answers directly to students less frequently than their counterparts. It could be argued that the pattern of T-S interaction of the individual teacher concerned is very much related to his/her beliefs about teaching and learning, about the role of practical work and about the nature of science. In particular, it is closely related to the teacher's expectation of how students should be involved in the lesson, as passive learners or as active learners, etc. The possible relationships amongst a teacher's pattern of T-S interaction and his/her beliefs about teaching and learning, about science and the role of practical work will be explored in more detail in the individual case stories in Chapters 4 to 11.

On the other hand, no appreciable trend was observed among the teachers regarding their tendency to give directions to their students and to offer assistance to their students during the TAS practicals. The same applies to how often they enquired about students' progress of work. The probable reasons underlying this difference have been discussed in an earlier section and will not be repeated here. It appeared, at least in Fay's case, that the high

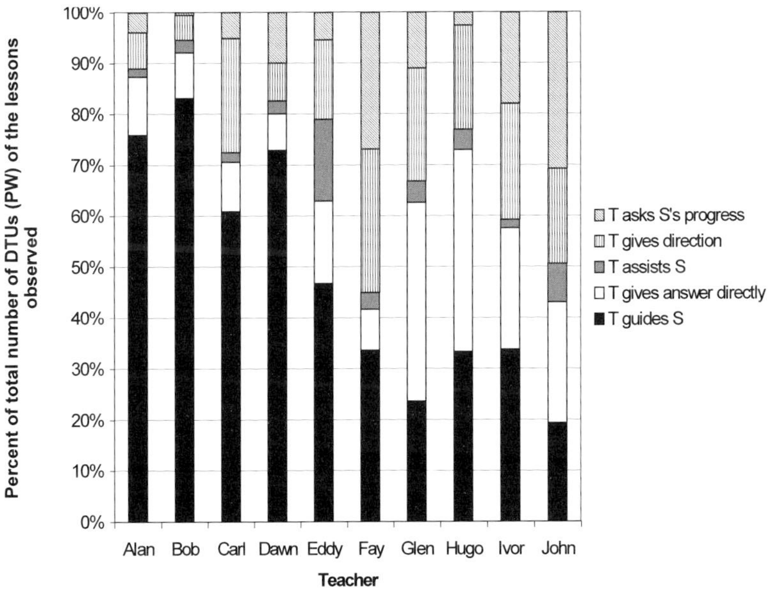

Figure 3-5. Patterns of T-S interaction related to teaching of practical work *per se*

proportion of time spent in this respect was due to the fact that the teacher was using this as a strategy to collect information for assessment purposes.

2.6 An overview of teachers' practices in the TAS

To conclude, teachers who were more concerned with assessment-related issues tended to interact less frequently with students during the practical, and hence less teaching related to practical work *per se* occurred. In particular, these teachers tended to spend less of their T-S interaction [in terms of % DTUs (PW)] on developing students' scientific thinking and understanding of the theory behind the practical work compared with the teachers who saw TAS as both a pedagogical reform and an assessment reform. Because of their concern with the issue of fairness, the more assessment-oriented teachers also tended to refuse to answer students' questions more often [in terms of % DTUs (PW)] than their counterparts. Interestingly though, they also tended to provide answers directly to students for the few questions that they did entertain. What are their underlying beliefs? This question leads us to the following chapters where I examine, case by case, the teachers' classroom actions in relation to their understanding and perceptions of the TAS, their beliefs about science, teaching and learning, as well as their educational visions and aspirations.

Chapter 4

ALAN

The students' companion

1. I'D NOT TEACH JUST FOR MAKING A LIVING

Alan was an idealistic and enthusiastic teacher in his mid thirties. He loved to develop analogies and metaphors for teaching and learning purposes. Before joining the teaching profession, Alan had a wide range of working experiences such as serving as a private practitioner, a house dental surgeon, an honorary research associate, and a part-time university lecturer. After the June 4th Incident in Beijing in 1989, he enrolled in a four-year full-time theology course during which he served as a part-time teacher in a secondary school teaching Katahelism and Biblical Knowledge. Upon graduation from the theology course, he taught Moral Education in a catholic secondary school for one year, at the end of which he left to join his present school, because he was not given a chance to teach biology. He started teaching A-level biology when he joined his present school in the year when TAS was first launched.

Alan enrolled in the theology course because he wanted to search for a meaning of life, which, he explained, was also related to his great interest in science ever since he was young:

> I liked science since I was a kid. I became very fond of it and was, in fact, superstitious of its magnificence in my secondary school days. When I was at university, I began to see its limitations, especially with what I saw and experienced when I was a research associate in the cancer clinic. A person can die of cancer without

B.H.W. Yung, (ed.), Assessment Reform in Science, 39-56.
© 2006 *Springer. Printed in the Netherlands.*

> any known reasons. I began to ask a lot more questions. Not on "How far can science develop?" but on "How much can it do for man and civilization?" This is the limitation which I am talking about. Hence, I switched over to look at life from another perspective, that is, from the level of meaning. Both biology and theology are related to life. Biology is more on the side of the hardware whereas theology is concerned with the meaning of life...

Alan regarded himself as an idealistic teacher, who "would not teach just for making a living". Instead, he wanted his students to learn something that would be useful for their entire lifetime – he said, for example, "learning problem solving skills is much more useful than learning the subject matter content". He also explained proudly, "Looking back on the students that I have taught, no matter whether they performed well in examinations or not, be they working in the Kodak photo finishing shop or as a policeman, when I talk to them, I know that they are surviving well in society." That was what Alan really liked about teaching, and, as he put it:

> Teaching can help in kids' development… It contributes to life. To their life. I really like to see life prosper… It is enough for me if my students can survive in society and become good citizens.

Alan's vision of educating his students to become good citizens was the reason for his switching to teaching biology after a year of teaching moral education and biblical knowledge because he realized that there was a problem of recognition. He found it difficult to get in touch with the students who treated him as a moral education teacher with a "saintly aurora" around him. He knew that it would not be effective to get his messages through via some form of dogmatic preaching. Thus, he wanted to achieve his goals via biology teaching, as he put it:

> It is much easier to get through to students via biology teaching. For example, when I say to them that all living things live for solving problems… This will then take me to topics on moral education and on questions about life. Joy does not necessarily come automatically. If you don't believe me, look at the tree, the fish… They are living in different ways, they are facing different problems. All living things are solving their own problems. Only when they can solve their problems, can they exist in this world. Hence, facing problems is the correct attitude for us in living life.

Alan had been trying hard "to achieve the dual goals at the same time". However, his school principal did not appreciate his efforts. During one of the interviews, Alan mentioned that he was considering leaving the school for a new environment (which he did at the end of the term):

> The learning atmosphere in this school is contrary to my philosophy of education... In this school, the students are too accustomed to being spoon-fed. The principal always complains to me that the pass rate for biology is always below average. So maybe I should just drill students to memorize and not bother if they understand or not. But I can't do this. So maybe I will quit. I am serious...

2. A SPIRAL OF TEACHING

In fact, Alan had been working very hard to adjust his teaching to suit his students. One of the characteristics of Alan's lessons was his frequent use of analogies, examples and illustrations from everyday life in explaining abstract concepts. For Alan, a successful teacher must be able to "go back to the students' level first in order to be able to bring them up". This was what he called "a spiral of teaching – brings it low; spirals it up". According to Alan, in order for "knowledge to be constructed", there has to be some groundwork or foundation. The foundation might be something taken from everyday life experiences, as Alan put it:

> I hope I can take them through via some everyday life examples. I'd try to make some cosmetic changes to the examples, to accommodate them, and then transform them gradually into the concept which I want to teach. In so doing, I hope that I can help removing from students the feeling of learning 'alien' ideas. That is, to make them feel that these things are quite simple, just similar to what they encounter in everyday life... In other words, carrying out an experiment is nothing different from something like their mother frying a fish in the kitchen (a metaphor which Alan made use of to explain the concepts of controlling and manipulating variables in carrying out an experiment).

Alan did not plan his lessons in great detail. His teaching style was interactive and he was in favour of "peer pedagogy". He characterized himself as a "dynamic" teacher who could teach in the classroom, in the lab, in the supermarket, in the countryside, using a book, or without using a book, with or without teaching aids. He would always look at his teaching from the students' perspective, which he attributed to the education course that he was attending.

> You have to accept your students as who they are. This is the basis of my philosophy of "bring it low, spiral it up"... My students are only of average abilities. They would learn only some but not all of what I teach them. I have begun to see that learning can be a process of gradual self-understanding undertaken by the students themselves. It is not necessarily a guided revelation, a sudden understanding guided by a tutor… I have begun to be more understanding of their situation. I can't say that I have lowered my expectations of them. I am just trying to look at the problem more from the students' perspective, from their abilities, their learning process…

3. PREDICTIVE CAPABILITY IS THE GREATEST THING ABOUT SCIENCE

Alan expressed a very articulated view of what science is and what its limitations are. He attributed that to his wide reading of books on philosophy of science while attending the theology course. His training at university also influenced him greatly in this respect. Though he is a Catholic, he didn't see any contradiction between his religious beliefs and his views about the nature of science. He believed that this would help his students recognize his open mindedness, his view about the nature of science, and his readiness to face challenges. The purpose of science, in Alan's view, was to seek to understand nature. He regarded predictive capability as the greatest thing about science. Thus, he did not regard the so-called social sciences as science because the objects they studied were not "stable, ultraistic, existential facts" as those in the study of life, the study of changes in matter, etc., are. In his view, "the application of a scientific method itself, does not necessarily mean that the discipline is scientific".

Allied to his religious belief, Alan cautioned about the danger of believing that "science is all mighty" and that "science has no limitations". In his teaching, he emphasized developing in students a realistic attitude towards science, as he put it:

> I want to cultivate in students an attitude towards science or even the same attitude in living as a human being. That is, to have a pragmatic attitude. The scientific quest is a never-ending business. This is because there is no perfect experiment. You can always find a loophole in any experimental design. When you try to cover up a loophole, some other loopholes will emerge... However, you can tell others proudly that you have tried your best to avoid that from happening. Hopefully, the error you introduced may be smaller than the error that you have eliminated. Whether this is true or not, you can't tell, you have to verify it again. Hence, this approach encourages them to have a pragmatic attitude in doing science and not to have unrealistic hopes.

Parallel to Alan's emphasis on cultivating in students a pragmatic attitude towards science was his emphasis on the attitude of objectivity in doing science. To Alan, "objectivity is the very essence that keeps science going". Thus, he highly valued those practicals in which the results are unexpected because students could learn to "pay respect to the facts objectively". He asserted that "very often it is this kind of side-track observations which become the sources of discovery..." Thus, the goal of science teaching, for Alan, was:

> The content to learn is just a tool. The most successful scientific learning is through using this subject knowledge, be it biology or others, to master the essence of science. That is, how to make knowledge? How to think? How to learn? That is the most important thing. That is, to cultivate in students a thinking habit. This will help them solving problems for themselves.

4. MIND-TRAINING THROUGH PRACTICAL TEACHING

Regarding the role of practical work in science teaching, Alan saw its importance as providing opportunities for students: (1) to develop their conceptual understanding, (2) to rectify their misconceptions, and (3) to practise problem solving. However, he didn't place too much weight on its role in verifying theories/concepts and in arousing students' interest. He believed that good quality practical work should leave room for students themselves to think. In his words, he aimed at "achieving the overall aim of mind-training through practical teaching". He hoped that, through the two years of A-level practical work, students could somehow master the investigative method, which later could evolve into some problem-solving skills that would be useful to them in the future.

Indeed, Alan had put a lot of effort into his teaching of practical work. He spent five lessons on what he called "theory of the practical" to prepare students for the hands-on practicals. He was so committed that he thought what he was doing ought to be included in the syllabus proper if the TAS was to achieve what it was advocating – approaching a problem methodically and systematically and then carrying out the practical effectively. He considered this a much better approach than asking the students to do many practicals, after which they didn't know much about the practicals at all. Indeed, Alan not only verbalized his understanding about the nature of science to his students, but also modelled it in his practice, as reported below.

5. CHARACTERISTIC FEATURES OF PRACTICAL WORK TEACHING

Of the ten teachers in this study, Alan was the one who interacted with students most during the TAS practicals. But he was also the teacher who touched on assessment-related issues least during the TAS practical. Of his T-S interactions on practical work teaching, 50% was focused on concepts underlying scientific research, 27% on the procedural aspects of the practical, 18% on biology theories underlying the practical, and about 2% on scientific attitudes. He saw the developing in students of "a mode of (scientific) thinking" and of appropriate scientific attitudes as the two most important goals of practical work teaching. Furthermore, he thought that practical work was also an avenue for students to learn to cooperate with each other, and to acquire some social skills that are important for their 'survival' in society.

5.1 Developing a thinking habit

Alan emphasized the cultivating in students of "a mode of (scientific) thinking" through the teaching of practical work. He designed and implemented a module of five lessons on what he called the "theory of the practical", before actually engaging students with the hands-on practical work. In these lessons, in the course of introducing to students basic concepts like 'hypothesis', 'control', 'dependent variable and independent variable', 'accuracy' etc., he often talked explicitly about the nature of science (see below). His emphasis on the importance of students acquiring a "habit of thinking" is revealed in the following excerpt from one of these five lessons.

TWC: What are the dependent variables, independent variables, objectives, etc.? This is what you have to develop in yourself as a kind of thinking habit. Okay? A thinking habit for yourself... Carrying out an experiment is relatively easier [than designing one]. The most important thing is that you must understand what you are doing. I reiterate once more ... the emphasis is not on the results, but on your mental skills. Okay? ... The process of your thinking...

Alan gave the same emphasis on this mind-training purpose during the hands-on practicals, as illustrated by the following episode.

Episode 4-1 – This is just a mind-training exercise
It was a practical on determining the water potential of potato tissue. A student was asking Alan for help in interpreting an unexpected result:
S: There is a decrease in weight! [instead of an increase as expected]
T: Didn't you weigh the strips while the potato strips were still wet?
S: Was I supposed to blot them dry first?
T: Yes.
S: In that case, water would be extracted out too!
T: This is also true.
S: What should I do then?
T: I have told you already that a needle cannot be sharp at both ends. You just comment on your method. There is no perfect method. This is just a mind-training exercise to find out what minute details you would consider in solving a problem. Right?

5.2 Developing scientific attitudes

Alan also emphasized the developing of appropriate scientific attitudes in students, in particular, an objective attitude in assessing experimental results. In order to ease students' worries if marks would be deducted from them for not getting the "expected results", Alan took the trouble to refer students to the relevant part of the TAS annual report.

TWC: The fourth line of the third paragraph, "Zero mark should not be given only because students fail to get the expected result." So, it doesn't matter if you cannot obtain the expected result. Just report your result. Report what you see. Report what you have done. Report what you think. (One student chipped in saying that "Should we report what we have broken?" and the whole class burst into laughter.)

Alan's emphasis on this goal was reflected in his insistence on this aspect when discussing with students during the practicals. A representative episode follows.

Episode 4-2 – An experimental result is an experimental result

S: Sir, shouldn't the result in the setup for xylene be the same as that for chloroform?

T: There is nothing like should or should not. An experimental result is an experimental result.

S: But I really don't know whether it is right or wrong. I [...]

T: You are carrying out an investigation, not solving a mathematical problem. There is a preset answer for a mathematical problem. But in an investigation, whatever you get is your result. What you should do now is to interpret and comment on your result. To find out if your result is reasonable or not...

In a similar vein, Alan reiterated the point that "there is no right or wrong answer" at the end of a post-lab briefing:

TWC: There is no right or wrong conclusion for this investigation. I am just testing your interpretation. How capable you are in applying your biological knowledge to interpret the data you've got. There are no right or wrong answers...

5.3 Developing social skills

After experiencing a case where students were fighting over some stock reagents on the teacher's bench, Alan decided to practise a new way of distributing chemical reagents – on a per bench basis. He hoped that through such an arrangement, students would learn to cooperate with one another, as revealed in the following excerpt where he was having a pre-lab briefing with his students.

TWC: Each of you should have a test-tube full of 90% alcohol. I have put on each bench a bottle of alcohol. Share it out among yourselves. I have given you more than enough. If there is not enough, it is absolutely your own responsibility to solve the problem. This is to test your cooperation – bench cooperation.

6. ASPECTS OF THE NATURE OF SCIENCE EMPHASIZED

Alan often talked about the nature of science explicitly in his lessons, especially during the five lessons on what he called the "theory of the practical". The following are three representative excerpts from these lessons.

TWC: Bear in mind that there is no perfect experimental design. At least, I haven't seen one. There are limitations for every experiment. Okay? When an experiment is finished, it is usually not the end. Usually it brings with it a lot of other questions which, in turn, lead to several hundreds of experiments. Okay? Therefore, there is never an end for experimentation. Never an end for experimentation.

TWC: One experiment may trigger off ten or even more experiments. This is science. It is directive. After one experiment, something will follow. This is how science develops – like building up a brick wall, one by one…"

TWC: The most important is repeatability and reproducibility. Any scientific experiment and any scientific evidence must be repeatable. If not, it is useless…

Alan's emphases on the ever-lasting nature of the scientific enterprise, its repeatability and reproducibility and hence its predictive capability, the imperfect nature of experimentation, and the importance of asking good questions, etc. were also characteristic of his discussions with the students during the practical sessions (see Episodes 4-1 to 4-8).

7. STUDENTS' ROLE IN THE PROCESS OF LEARNING

In line with his goal of developing in students "a thinking habit", one characteristic feature of Alan's classroom practice was the ample opportunities he created for his students to get actively involved in decision making in the learning process itself. For example,

- providing students with a standard set of equipment at the beginning of the term, such that they have to decide for themselves what to use in a particular experiment in relation to their own experimental designs
- grasping every chance to encourage students to pursue further investigations that spin off during the course of work (see Episode 4-5)
- brainstorming with students to stimulate them to develop and to formulate for themselves the aim of the experiment that they are going to perform

Alan's intention to encourage students' active participation in the learning process was reflected in the following representative excerpts from his lessons.

TWC: There is no one specific method. There can be many different ways to solve the problem. It is like the Eight Fairies crossing a river [each in their own unique way]. Really, this is to crack your brain.

TWC: ... Learning by doing. It is useless to tell you the theory of how to ride on a bicycle. You must ride on a bicycle to learn how to do it.

T: Never mind, learning by doing. Making mistakes is a good thing. Carry on with your work…

As evident in the last two excerpts, Alan strongly believed in "learning by doing" and "learning from mistakes". In a similar vein, Alan allowed the greatest freedom for his students to carry out the practical according to their own designs, despite some of them not being as good as he had anticipated. He preferred to let students carry things out first and then bring up problematic features of design for discussion during the post-lab discussion. The following is one such representative episode from a practical on investigating the factors affecting membrane permeability of beetroot tissue.

Episode 4-3 – Are you preparing 'minced' beetroot?

T: Are you preparing 'minced' beetroot? (Alan noticed that, in so doing, it might be difficult for the student to control the variable of surface area for diffusion of the red pigments if the beetroot tissue were to be chopped into fine pieces like minced beef.)
S: This can increase the surface area for the red pigments to diffuse out.
T: Oh! I see! I see what you mean. I see what you mean. I see. I see. I see, I see.
T: Are you going to repeat this for all the cylinders?
S: Yes, more or less.
T: Okay. Okay.

Alan also wanted his students to persevere and to be active in seeking ways to solve the problems that they encountered in the process of learning. Instead of telling them the answers directly, he preferred to guide students through to the answers, as revealed in the following episode.

Episode 4-4 – Having problems is not a problem

S: Sir, will this one become different after immersing it in water?
T: Yes, you raise a very good question. What should you do then?
S: What should we do then?
T: Think about it yourself. What should you do then?
S: [...]
T: No matter which method you use, each will have its own problems. What is the problem with this method?
T: Having problems is not a problem. Think of a method to solve it.

Another method employed by Alan to encourage students to be active learners was prompting them to reflect on, and explore, possible interpretations of their findings, as revealed in the following episode.

Episode 4-5 – I can tell you frankly that I don't know why either

T: Did you notice that this particular piece of beetroot cylinder floats, whereas others sink? What do you think?
S: Maybe it is dead.
T: Why is it that it will float when it is dead?
S: No, maybe some air has gone into it.
T: So, there can be many possible reasons. And your suggestion is that it has nothing to do with the cells?
S: Maybe you're right. Maybe this is due to inherent characteristics of the cells.
T: Why is it that all others have no air bubbles except this one?
S: Maybe this one is rotten already.

T: Could it be that? Is there any way that you can gather more evidence to support your hypothesis?

The discussion continued for some time and was then wrapped up by Alan:

T: Yes. Go and test your idea with those outside. (The teacher is pointing to the left over beetroot cylinders on the teacher's bench.) This is collection of evidence. You shouldn't say things casually.

Some time later, Alan returned to the student and resumed discussion with him:

S: Sir, I got the evidence now.

T: What evidence have you got? Show me.

S: There are many white spots in this beetroot cylinder too. These are the less dense materials.

T: Um! Um!

S: You see. When you compare it with this one, there are fewer of these [white spots]. Hence I think it is something to do with these white spots.

T: Um! Um! That means, you think that the beetroot cylinder floats because of these white spots.

S: Yes. Maybe these [the white dots of tissues] are less dense and hence it floats.

T: Is this white in colour too? (The teacher is pointing to another piece of beetroot which sinks.)

S: In that case, this (the presence of white dots) is not supporting evidence.

T: Retain this sample for me... I can tell you frankly that I don't know why either. That's why I asked you to retain the specimen for me. I don't know why it floats. I haven't seen that before either. So, after you have finished, throw away all but that particular tube. Okay? I'll put it in the refrigerator and investigate further when I am free tomorrow.

Overall, there was a lot of T-S interaction taking place during the lessons. Most of the time, Alan was guiding students (76%) instead of giving them the answers directly (11%). In other words, the students had to construct their own knowledge under the guidance of the teacher.

8. CHARACTERISTIC FEATURES OF ASSESSMENT PRACTICES

Alan required his students to complete their lab reports in class. He set the time limit for handing in the report according to the progress of work of his students, as illustrated in the following episode.

Episode 4-6 – That's a deal

S: Sir, when do we have to hand in the report?

T: Of course, at 3:15 p.m. That is my tea time (when school ends).

S: How about 3:30 p.m.?

T: Okay. That's a deal. Hand in your report at 3:30 p.m.

However Alan did not enforce that deadline stringently. Instead, he had an allowance of 30 minutes for those slow workers to finish writing up their reports after the time limit.

As revealed in the previous episodes, Alan did not avoid discussion with individual students. In fact, he often initiated discussions with individual students. On the issue of whether to allow students to discuss among themselves during the practicals, Alan took a rather relaxed stance, as revealed in the following two episodes.

Episode 4-7 – You carry on with your discussion

T: Why don't you ask me [instead of asking your neighbour]?

S: We discuss it because we are classmates and we can learn from each other.

T: In that case, carry on with your discussion. I won't interrupt you then. You carry on with your discussion.

Episode 4-8 – Have a look at somebody else's results first

S: Based on this set of data, can I draw a straight line here [for the graph]?

T: The way you draw your graph will affect your conclusion. You decide yourself. Maybe, you could go to have a look at somebody else's results first. Because whether you draw a straight line or a curve, this would affect your final conclusion.

T: Go and look at Tom's, Greg's and Steve's results before you decide on whether to draw the best straight line through or to draw a curve.

Overall, assessment activities did not constitute a prominent feature of Alan's practical work instruction. In other words, Alan was able to integrate assessment with his teaching of practical work quite successfully.

9. BELIEFS UNDERLYING CLASSROOM PRACTICES

In order to explore the possible relationships between Alan's classroom practices and his underlying beliefs, it is essential to examine his underlying beliefs in relation to his classroom actions. For the sake of presentational clarity and ease of comparison with other cases, beliefs underlying (1) his teaching of practical work *per se* and (2) his assessment actions will be examined one after another.

9.1 Beliefs underlying teaching of practical work *per se*

9.1.1 Knowledge has to be constructed by students themselves

As evident from the episodes and the portrait presented above, even a relatively casual observer is likely to notice the close relationship between Alan's classroom practices and his espoused beliefs. In particular, there was a close correspondence between his conscientious efforts in trying to get students actively involved in the learning process and his beliefs that "knowledge has to be constructed" by the students themselves. In addition to those instances reported above, the following interview excerpts provide further evidence to support this.

> Nowadays, the worst thing is that students have lost the interest to find out the answers for themselves because they have become accustomed to being spoon-fed. Everything is fed to them including the problem to be investigated. But this is contrary to the goal of science education.

> If we over-prepare things for students, we would be quenching their imagination. This would make them very passive... As long as it is not too dangerous, I have no problem letting students work according to their own proposals. It doesn't matter even if their method is different from mine...

> I always say to my students, "Can I predict in advance all their questions and tell them all the answers beforehand?" If not, then they have to answer the questions by themselves. That is, they need to have the ability to evaluate their own actions, their own decisions. This is important to everyone be he a layman or a scientist. Therefore, I would provide as many opportunities as possible to exercise their brain. I hope I can cultivate in them some intellectual hobby, habit or mode of thinking. I am aiming at achieving the overall aim of mind-training through practical work teaching.

9.1.2 Developing students' scientific attitude

Alan's personal interest in the philosophy of science and his experience as a research associate also had a great influence on his teaching of practical work. Besides those noted above, his beliefs in the importance of cultivating in students an objective attitude in doing science was notable, as revealed in the following interview excerpt.

> Developing a scientific attitude in students is more important than teaching them the scientific method. Because method is just like a knife. The knife can be used to kill a person or to save a life. Some people practise the scientific method, but they use it in a

> very subjective manner. They may be in the 'shape' but not in the 'spirit'. Thus, I think acquiring the scientific attitude is more important.

9.1.3 Developing some lifelong problem solving skills

Clearly, Alan had clear and strongly held beliefs about the purposes of science education that were akin to those advocated by the TAS. This explains why he did not mind spending 20 minutes on brainstorming with students about the possible investigations that could be undertaken based on a certain observation – an action which other teachers might have seen as not worthwhile. But Alan asserted that:

> It depends on what you are looking for. If I can do something which students can remember well in their future, that is, they learn how to use a certain method to investigate into something, I think it is worthwhile from this perspective. Of course, it is not worthwhile if you look at it from the perspective of submitting marks to the public examination authority or from the perspective of rushing to finish the [examination] syllabus.

It is this last sentence that distinguishes Alan from those teachers who regarded TAS primarily as an assessment reform. He believed in the long-term goal of developing in students some lifelong problem solving skills rather than the short-term goal of preparing students for the examination, as revealed in the following interview excerpts.

> I hope that during the two years of practical work, students can somehow master the investigative method, which later can evolve into some problem solving skills that are useful to them in the future. I don't want them to do something just purely for going through the motions.

> I deliberately distribute the materials on a per bench basis. It is better to ask them to settle disputes face to face among themselves. I want them to learn to cooperate among themselves. This is a skill that is required in real life. This is just the same as coping with stress, coping with examinations, etc. These are some essential skills for survival in a civilized way...

Clearly, in the course of teaching practical work, Alan had drawn into his mind other considerations that were not directly related to the teaching of specific content. Indeed, he was always aware of his goal of "educating students to becoming good citizens". To sum up, Alan did not just welcome the introduction of the investigative approach, as advocated by the TAS. He actually invested a lot of effort in making it work for his students, whom, he hoped, would develop some lifelong problem solving skills through practising the investigative approach. This was in line with (1) his beliefs about the nature of science, (2) his beliefs that "knowledge has to be constructed" by students themselves, and (3) his vision of "educating a good citizen".

9.2 Beliefs underlying assessment practices

9.2.1 I see my role as a students' companion

As mentioned earlier, assessment activities did not constitute a prominent feature of Alan's practical work instruction. This could be explained by how he saw his role as a teacher in the context of the TAS.

> I would never tell students on the spot that I have deducted marks from them. Under these circumstances, from the point of view of the student, the teacher is a spokesman of the exam board. He is the adjudicator. But I see my role as a students' companion. I am with them to cross over the hurdle of the TAS.

Thus, it is understandable why he took a rather relaxed stance on the issue of whether to allow students to discuss among themselves during the TAS practicals. Underlying his practice were also his beliefs about teaching and learning as well as those about the nature of science. As he put it,

> Discussion is very important to learning, the development of ideas, the development of science as a whole. I encourage students to do so as part of peer pedagogy. This is to let them behave like scientists…

On similar grounds, Alan did not avoid giving assistance to students during the TAS practicals if he deemed it necessary. He upheld his beliefs that the teacher's role was to assist students' learning, and that the TAS assessment requirements and regulations should only be of secondary consideration in the process, as revealed in the following interview excerpt.

> On practical grounds, I would have to give them assistance. If not, I don't know what they would come up with… If such circumstances do arise, you still have to do so.

For Alan, discussing with students and/or giving assistance to them simply meant "guiding" and "focusing" instead of "telling and giving the answer", which is what it would mean for many other teachers in this study. This explains why Alan did not mind discussions with students, unlike several of the teachers. This is because Alan believed that there was still "a long way" for students to "struggle through" before they could come up with a decent report, and thus discussing with them would not affect the fairness of the assessment to a significant extent. Nevertheless, Alan was still aware of the issue of fairness, as revealed in the following two interview excerpts.

> It is impossible for me to discuss with each of them individually. I do it all in one go by putting it on air, broadcasting it to the whole class. This is for fairness sake.

> I would require all pupils to hand in their reports within 30 minutes after the first student has handed in the report. This is purely for the sake of fairness. It is a deadline. But whether it is a stringent one or not depends on how you look at it.

9.2.2 Integrating assessment with teaching and learning

Overall, Alan's assessment actions were very much related to his pedagogical beliefs, which were mostly learning-focused; that is, focusing on facilitating students' learning and understanding rather than teaching *per se*. Thus, he would adjust his assessment practices to accommodate and integrate them into the teaching-learning situation in such a way that they could best facilitate students' learning. Alan's response to why he used impression marking, rather than checklist marking in assessing his students, was representative of his views on this aspect.

> Since I allow them a good degree of latitude in designing their own experiment, I prefer impression marking. My students are just too brilliant. It is just impossible for me to predict what they will come up with regarding the experimental design.

The following excerpt further reveal Alan's understanding of the TAS also as a pedagogical reform which advocates the teaching of practical work using the investigative approach.

> Without the TAS, I might not have organized the so-called "theory of the practical" into some sort of a mini module spanning five lessons. Certainly, the TAS requirement to carry out investigations is a catalyst. The TAS is the driving force. I think the TAS is a revolution. No, not exactly, but the investigative approach it advocates... Teaching is a long process of struggling... My feeling is that I am improving year after year.

Also evident from the above excerpt is Alan's sense of professionalism and how it is related to his reaction to the TAS reform.

10. AN OVERVIEW OF ALAN'S CASE

In brief, Alan was an idealistic and enthusiastic teacher with a vision of education beyond the classroom – educating students to be good citizens and cultivating in them "a thinking habit". He had a strong view of what science is. He emphasized the cultivating of a 'pragmatic' attitude (i.e. realizing the limitations of science) in his students, and making them realize that science is a never-ending business. His beliefs about teaching and learning were mostly learning-focused (i.e. to be concerned if students can really learn what they are taught). Most notable was his teaching philosophy of "bring it low; spiral it up". To him, content is just a tool with which to learn the method of how to solve problems.

In all, there were close matches between Alan's practice in the teaching of practical work and his beliefs about science, teaching and learning. In particular, his emphasis on developing in students appropriate scientific attitudes was very much related to his personal beliefs in the philosophy of science – notably, amongst others, his beliefs about the imperfect nature of experimentation, the ever-lasting nature of the scientific enterprise, its repeatability and reproducibility and hence its predictive capability, constituted major features of his discussions with the students. His conscientious efforts to provide students with ample opportunities to get actively involved in the learning process tallied with his belief in the goal of cultivating in students "a thinking habit" and "some problem solving skills that will be useful to them in the future". His vision of educating students to become useful citizens was also a focal aspect in his teaching. This was reflected in his practice of distributing chemical reagents on a per bench basis so that students had to learn to cooperate with each other. Above all, his understanding of the TAS as both an assessment reform and a pedagogical reform was central in determining what he would bring forth to figure in his focal awareness during his teaching of practical work in the context of the TAS, and what to recede to the background. Clearly, in this way, his interpretation of the nature of the TAS as a revolution and a reform was closely related to his beliefs in science and in the goals of education.

Chapter 5

BOB
The teacher with a mission

1. I'M A TEACHER WITH A MISSION

Bob was an energetic, committed teacher in his late twenties. He had wanted to become a teacher when he was still an undergraduate. Teachers he had known, from those in his primary education to those in his teacher education course at university, were important influences in his choice to become a teacher.

Bob was vibrant and enthusiastic, consumed with his search for what it would take to be a good science teacher, as revealed in the following excerpt.

> I am a teacher with a mission. I hope I can improve continuously. I ask my students to evaluate my teaching nearly every year. At the end of each academic year, I ask myself several questions: Why do I have to teach biology? What is good science teaching? Why is the learning of scientific method important? … Once I go into the Education Library, I always have a dream. How wonderful it would be if I didn't have to go to work, but to study the books there for half a year because there are so many things that I don't know, those teaching methods, etc. I really feel that way.

Bob did not restrict his role of being a teacher to only the teaching of subject matter; rather, he included being a teacher "who cares about the

B.H.W. Yung, (ed.), Assessment Reform in Science, 57-67.
© 2006 *Springer. Printed in the Netherlands.*

students in all aspects of their development". He regarded himself as "a very suitable person to make friends with young people".

2. GOOD TEACHING IS GOOD LEARNING

Bob was not the type of person who would give in to social norms easily by "choosing the easy way out". He would stand firm on things which he thought were beneficial to his students.

> It has been an headache for me to negotiate with my colleagues on how the Form Three [i.e. Year 9] examination paper should be set… I strongly believe that asking students to memorize all those terms would just scare them away from learning biology. I was very upset. I kept raising this point every time when we were about to set the paper.

Indeed, he grasped every possible chance to make use of the flexibility provided by the system to do something that he thought would be beneficial to his students' learning.

> Teaching at Form Six [i.e. Year 12] is the least constraining for me because I can teach in whatever way I like. I am the only person who decides what to examine. I can even skip one of the regular tests [required by the school authority] and replace it by the individual project which, I think, contributes significantly to students' learning.

As revealed in the two excerpts of interviews above as well as the one below, Bob held beliefs that were mostly learning-focused rather than teaching-focused.

> Good teaching is good learning. Much of the learning has to be carried out by the students themselves. The teacher is just a facilitator to inspire students to think along certain directions... The more they think, the more they learn…

3. SCIENCE IS TO FIND OUT THE TRUTH

Bob liked to read books about the philosophy of science. He regarded scientific method as central in the development of science. In his view, scientific ways of thinking could be applied in a wide array of situations, not necessarily restricted to those within the scientific arena. He also understood that sometimes scientists just did not work in ways that are generally referred to as the scientific method. He also believed that "observation is theory laden"; as he put it:

> I think everything is backed up by a theory. Even with something as simple as a measurement, there must be a number of preset conditions before one can carry it out. In the past, I thought that science was a very objective way of doing things. Now, I find that it is not as simple as that. A hypothesis may only be the best hypothesis at a particular time. It could be changed when there are new findings.

Though believing that science is not static, but dynamic and on-going, Bob – being a Christian – maintained that "there is Truth. However science is not able to arrive at it even though the spirit of science is to find out the truth." In his words, "That involves a leap of faith."

Teaching science, for Bob, comprised three levels, "It can be teaching some simple facts like teaching ABC as the ground work for learning English, without which students cannot proceed further. The next level involves understanding the relationship between the parts and the underlying reasons and concepts. The highest level is the method of thinking – the thinking underlying scientific method. Students have to know the assumptions, observations, results and limitations behind a theory. They have to be engaged in some processes of thinking. They have to ask the question of why? Just learning a certain psychomotor skill is not learning science at all. "

4. THE ROLE OF PRACTICAL WORK

Regarding the role of practical work in science teaching, Bob saw its contribution in: (1) inculcating in students a scientific attitude, (2) enhancing students' understanding of scientific method, (3) enhancing students' problem solving abilities. In his view, "Problem solving is a more important goal than verifying theories and concepts. Students are not going to get a huge conceptual understanding through practical work." On top of these,

Bob saw additional advantages in asking students to carry out an individual project of investigative practical work. These included stimulating students' creativity, developing their communication skills and cultivating in them a sense of achievement.

5. CHARACTERISTIC FEATURES OF PRACTICAL WORK TEACHING

The amount of Bob's interactions with his students during the TAS practicals was the second highest of the teachers in this study. He was second least in his attention to assessment-related issues during the TAS practicals. Of the T-S interactions related to practical work teaching, 53% was focused on concepts underlying scientific research, 25% on biology theories underlying the practical, 21% on the procedural aspects of the practical and about 2% on scientific attitudes. Most of the time, he was guiding students to the answer (83%), rather than telling them the answer directly (9%).

5.1 Enhancing understanding of the concepts underlying scientific research

Bob was one of the two teachers who asked students to carry out an individual project of an investigative nature. Actually, he had practised this even before the time of the TAS. But at that time the result was not satisfactory. According to Bob, the students were not enthusiastic about it because they knew that it did not count towards the examination marks. The situation had improved a lot after Bob turned the project into one of the requirements of the TAS; that is, through valuing the project as part of the assessment regime in the TAS, Bob hoped to "bait" students to invest more effort into it.

> Through the project, they can experience the kind of joy that they can get out of it and have a more thorough understanding of the scientific process and the skills involved... The TAS provides me with justifications for asking students to put more effort into it ...

Indeed, Bob strongly emphasized developing in students an understanding of the concepts underlying scientific research. He often initiated discussions with individual students on the assumptions and limitations underlying their experimental designs. The following are two representative episodes in this area.

Episode 5-1 – Identify all your limitations

TWC: This is going to be a simple practical, so try to be more critical in your report. So identify all the assumptions, identify all your limitations, etc.

Episode 5-2 – Why do you think that you can assume …

It was a practical on determining the water potential of potato tissue. Bob was discussing with a student on how much sucrose solution should be used for immersing the potato tissue:

T: So you are using 5 ml of sucrose solution for the experiment, right? Why do you choose to use 5 ml but not other?

S: It is easy to make up this concentration [using the stock solution provided]. In addition, since there will be water moving in and out of the potato cells, it won't lead to a change of concentration in a 5 ml bathing solution.

T: UhUh.

S: I think it is easier to make up the solutions. Also, it won't waste any material.

T: Why do you think that you can assume that there will be no change in concentration in the external bathing solution?

S: Because I think there is too little [water movement in and out of the potato cells] to make any changes.

T: So this is your assumption?

S: Yes.

T: What do you think if you want to make sure that your assumption is a valid one; that is, there is no change in the concentration of the bathing solution outside? Given that you understand all the limitations, given the amount of sucrose solution and materials provided. You are provided with quite a lot of sucrose here. What do you think you should do?

S: You mean?

T: What should you do in order to be sure that your assumption is valid as far as possible, given the limited materials and the limited time for carrying out the experiment?

S: How can I prove it?

T: You can't prove it – it is your assumption. How can you be sure that this is a valid assumption; that is, there will not be any change in concentration of the bathing solution, or the change will not a significant one?

S: […]

T: This is related to the volume of bathing solution to be used.

S: […] (?)

T: No, no. You can't, you cannot detect the changes in the solution.

S: You mean how do I know?

T: What I mean is, your choice of the volume of immersing solution to be used is going to affect the validity of your *assumption* that there will be no change in concentration in the immersing solution.

S: That means the greater the volume the better it would be?

T: Yes, right!! In fact, I expect you to use 10 ml because I provide you with 40 ml. If you do some calculations you'll find out that if you use 10 ml, you can make full use of the materials provided.

5.2 Enhancing students' problem solving ability

Bob gave great importance to developing in his students problem solving skills through the teaching of practical work, as revealed in the following episode.

Episode 5-3 – I am just using it as some sort of problem solving exercise

TWC: Please get a handout here, and read the background information…
S1: Respiratory quotient?! What is it?
TWC: Don't worry about that. You'll know what's going on after you've read the background information.
S2: Sir, RQ [Respiratory quotient] is not in the syllabus.
TWC: Yes, I know that it is not in the syllabus, but I'd like you to work on that.
S2: Why?
TWC: It is used as a problem solving investigation.
S2: Oh! No. We don't want out of syllabus materials. We'll all get lost.
TWC: No panic. I am not going to teach RQ *per se*. I am just using it as some sort of problem solving exercise.

5.3 Developing scientific attitudes

Bob paid great attention to cultivating in students appropriate scientific attitudes. In particular, he did not like his students fabricating experimental results, as revealed in the following episode.

Episode 5-4 – If you are not sure … then you should not report this as a result

This was an experiment on determination of water potential of the epidermal cells of a plant. Bob was checking the epidermal preparation of a student through the microscope. He was suggesting to the student that he should be faithful to the experimental data:

T: Would you say that this [particular cell] is not plasmolysed? (of which the student has reported as plasmolysed)
S: Yes. But it could also be regarded as an inaccurate observation. (grin)
T: Come on, you are just cheating.
S: What cheating? It's an error [of observation], not cheating, you know.
T: But you know certainly that it is plasmolysed.
S: But you can't tell if this is really a cell or not, (?)
T: If you are not sure if it is plasmolysed or not, then you should not report this as a result.

6. ASPECT OF THE NATURE OF SCIENCE EMPHASIZED

In his discussions with individual students during the practical, Bob often talked about the nature of science, though in an implicit manner. In particular, he wanted his students to understand the imperfect nature of experimentation, as revealed in the following two representative episodes.

Episode 5-5 – You have to strike a balance between reliability and the time available

S: Sir, am I supposed to count all the cells in the field of view [of the microscope]?

T: What do you think?

S: [...]

T: In fact, you can decide on your own method of counting and recording. Of course, the greater the number of cells you count, the better it is. But you have to take into consideration the time available. You have to strike a balance between reliability and the time available.

Episode 5-6 – Any form of data treatment will lead to different kind of errors

It was a practical on determination of water potential of potato tissue. A student was not sure if it would be okay for him to estimate the water potential from a graph plotted from 6 readings.

S: Can I not report the exact value but just say that it is close to 0.4 M? Is it okay?

T: As you know, any form of data treatment will lead to different kind of errors. It is okay as far as you report on what you did and that you understand what are the problems that are associated with your data.

7. STUDENTS' ROLE IN THE PROCESS OF LEARNING

Bob wanted his students to actively participate in the lessons. In particular, he often initiated discussions with individual students, be they on experimental design or procedural matters. He wanted them to be able to rationalize about the methods they used and the procedure or actions that they had taken. If a student did not know an answer, instead of telling the student the answer directly, Bob always allowed the student some time to think about the question before he resumed the discussion with the student at a later stage. In his own words, Bob often wanted to engage his students "in some processes of thinking", as revealed in the following episodes.

Episode 5-7 – I have a question for you ... think about it first. I'll come back to you later.

T: I have a question for you. Now, you've put the peeled epidermis in water. Do you think this will affect your experimental results or not?

S: Oh yeah! It will. I should not have done that. (In fact, putting the epidermis in water will not affect the experimental results. Bob was just trying to find out if the student understood the underlying biological theory.)

T: Really? But in the last experiment, you didn't put the potato strips in water. But this time, you put the epidermis in water. I am not saying that this is wrong. But you may think about it first. I'll come back to you later.

Episode 5-8 – You have to consider the trade-off

T: Wait a minute.

S: A mistake again?

T: No, no, it's okay. But you have to consider the trade-off, if you cut the beetroot tissues into very thin slices, then it will be very troublesome later on.

S: Is it that it would be very difficult to weigh the tissues if I cut them into such thin slices?

T: Yes.

8. CHARACTERISTIC FEATURES OF ASSESSMENT PRACTICES

Bob assessed his students in nearly every practical. He distributed the lab manual to his students only when the practical was due to start. During the practical, he often initiated discussions with individual students. He adopted a tolerant attitude towards discussions among the students themselves as long as "they didn't stretch the limit too far". He required students to complete their reports in class. The deadline for submission of reports was determined according to the students' progress of work. There could be a difference of about 15 minutes to half-an-hour between the first and last student in handing in their reports. Overall, assessment activities did not constitute a prominent feature of Bob's practical work instruction. Rather, he was always aware of "sort of listening to and approving students' experimental designs" so as to give them confidence to carry on with their work, as revealed in the following representative episode.

Episode 5-9 – Yes... Let's see... What if... Well, it's up to you. You decide yourself.

S: To put it simply, we have to prepare solutions of different concentrations... Then measure the initial weight [of the potato discs]... Immerse them until an equilibrium is reached. At that moment you take the potato strips out, you have to blot them dry.

T: Yes.

S: But then will the contents inside the potato cell saps be drawn out too? I think this will introduce a great experimental error.

T: Let's see if this would produce a very large error or not. What if you don't place the tissue paper onto the potato strips for a very long time and you just try to blot dry the visible liquid on the surface?

T: Well, it's up to you. You decide yourself.

9. BELIEFS UNDERLYING CLASSROOM PRACTICES

9.1 Beliefs underlying teaching of practical work *per se*

As evident from these episodes and the portrait presented at the beginning of this chapter, there were close matches between Bob's classroom practices and his various educational beliefs. In particular, his usual practice of initiating discussions with individual students, and that of giving them some time to think instead of telling them the answer directly, were very much related to his belief that "the teacher is just a facilitator…, much of the learning has to be carried out by the students themselves" and that "good teaching is good learning".

Bob's practical work instruction was also very much related to his beliefs about the nature of science, especially his belief that "everything is backed up by a theory". That's why he often asked his students about the assumptions underlying their experimental designs. This was also in line with his view that "the highest level of science teaching is the method of thinking – the thinking underlying the scientific method… (such as the) assumptions … and limitations behind the theory."

Bob's full support of the investigative and problem solving approach advocated by the TAS was fuelled by his firm belief in the role of practical work in enhancing students' problem solving abilities. This is reflected in the following interview excerpt taken from his response to why he asked students to carry out a practical outside of the public examination syllabus.

> It is a problem-solving exercise. Students have not learnt what R.Q. [respiratory quotient] is. But based on what they have learnt, together with some background information provided in the lab manual, they should be able to work out a method to find out the R.Q... To practise problem solving through practical work is the most important. This is line with the demand of the TAS.

Bob's belief in the role of practical work as inculcating in students appropriate scientific attitudes was also evident in his teaching. In fact, he was quite satisfied with his achievement in this respect.

> In the past [when there was a public practical examination] students could fabricate the experimental results without actually carrying out the practical. No one could check this out. Students did not have a very good attitude in this respect… Now, the

> students understand my expectation that they should never fabricate results, and that they must record whatever data they get...

9.2 Beliefs underlying assessment practices

As mentioned above, assessment activities did not constitute a prominent feature of Bob's practical work instruction. He would not avoid interacting with students during the TAS practicals because he thought that he was just "sort of listening to and approving their experimental designs" and "there was no need to avoid discussion with individual students unless it came to the discussion and interpretation of results". Nonetheless, Bob was caught in a dilemma of whether he should allow his students to discuss among themselves.

> This is a dilemma for me... Students will learn more if they discuss with each other...In fact, this is a difficult problem for me... The TAS does not allow this to occur otherwise you will not be able to assess individual work... I can't think of any solution... It is still tolerable if a student asks his classmate on one or two points ...

That is, Bob was torn between facilitating students' learning and accomplishing the assessment requirements. In the end, he decided to adopt a tolerant stance on the issue for the sake of students' learning.

10. AN OVERVIEW OF BOB'S CASE

To sum up, Bob was committed to "improve himself continually". He often strove for things that were to the benefit of the students. His vision of education also extended beyond that of the classroom because he "cares about the students in all aspects of their development". His belief in the role of the teacher as a facilitator of students' learning and his belief in the importance of the thinking process coincided with his belief in developing students' problem solving abilities as a more important goal in practical work teaching. All in all, Bob held beliefs that were mostly learning-focused rather than teaching-focused. Bob also had an articulated view about the nature of science and the role of practical work in school science education. In particular, he believed that "everything is backed up by a theory" and that science is dynamic and on-going.

Bob emphasized enhancing students' understanding of the concepts underlying scientific research, developing their problem solving skills as well as inculcating in them appropriate scientific attitudes. All these tallied with his beliefs about the nature of science, in particular, that "everything is backed up by a theory" and that "the highest level of science teaching is the method of thinking – the thinking underlying the scientific method". Also evident from his practice were his beliefs in the role of teacher as a facilitator of students' learning and that "good teaching is good learning". Above all, his understanding of the TAS both as an assessment reform and a pedagogical reform was central in determining what he would bring forth to figure in his focal awareness and what would recede to the background during his teaching of practical work in the context of the TAS. On the whole he was pedagogically-oriented rather than assessment-oriented when faced with the dilemma of facilitating students' learning or accomplishing the assessment requirements.

Chapter 6

CARL

The teacher committed to all-round education

1. CREATING "INVITATIONS TO LEARNING"

Carl was a very experienced and innovative teacher, who was extremely gifted in the art of communicating his thoughts. In his first year of teaching, he served as a laboratory teaching consultant, teaching only A-level biology practical classes. He taught Certificate-Level Biology for another two years in another school before going to England to study for an Advanced Diploma in Education course. On his return, he joined the present school and has taught Advanced-Level biology since then. During this period, he obtained his second teacher education qualification and Master in Education. Perhaps because he had studied three education courses, Carl was able to articulate his vision of education very well.

> Nowadays, we need students with more creative minds, … to feel themselves playing a major role in history… Maybe, I am not looking for to cultivate genius but to develop citizens who can actively solve their problems. I think science education can achieve this goal. I am not expecting any immediate results, but it is important to sow the seeds and to have something rooted in the students.

Indeed, Carl had been encouraging and supporting his students to participate in the joint school science exhibition. He did not like to drill students for examinations because he regarded it as a violation of his belief

B.H.W. Yung, (ed.), Assessment Reform in Science, 69-85.
© 2006 *Springer. Printed in the Netherlands.*

in all-round education. Instead, he often tried to "plan an integrated learning experience" for his students by linking up biological knowledge with some other seemingly unrelated things. He saw the value of this in terms of training students "to look at things from different perspectives" and showing to them "the existence of certain resources". This latter notion of "awareness of the existence of certain resources" was in line with Carl's idea of liberal education, lifelong education and his vision of the future society.

> The prerequisite for learning is students' motivation. The teacher has to create many learning contexts, so-called invitations to learning. Once the students accept one of these opportunities, they will pick it up and explore it further on their own. However, what is the exact 'button' that can 'turn on' their learning in a certain area? No one can tell this in advance. In my view, a healthy society should be able to provide something positive to its citizens via many different channels, hoping that they can pick up some of the good aspects and carry on their own lifelong learning. This is my vision of future society... Learning takes place not only in the school. There are a lot more learning opportunities in the family situation, be they from the media or from the parents or from their visits to other places...

2. ACTIVE LEARNING

Carl's idea of lifelong learning was in line with his idea of the importance of active learning on the part of the learner; as he put it:

> I accept that the human mind is an always-changing entity. So humans are processing new information continually, to replace, to deconstruct and to change previous conceptions.

Accordingly, he designed his teaching materials to be more "student-participatory" so as "to help them to make cross comparisons, to extract the main points, to link up previous knowledge with the newly learnt knowledge". In Carl's words, "this is what real learning should be. A student with real learning has a systematic structure with which to organize the information which he has to 'memorize'." Interestingly, Carl viewed the role of memorization in learning in the following way.

> Memorization is a necessary evil. However it occurs not through rote memory but through repeated use… I don't want to kill students' interests just because the topic requires a lot of memorization… We should integrate the concepts such that the students would have memorized something unconsciously… We should develop strategies which enable students to process the same dry facts from different angles so as to facilitate their memory. Multimedia teaching may have a part to play in this respect.

Indeed, Carl used a lot more audiovisual aids in his teaching of practical work than other teachers in this study. Carl also spent relatively less time on marking assignments. Rather, he spent more time on lesson preparation and interacting with students during the lesson, because he preferred "to deal with people face to face, to make use of the teaching situation to guide students to progress step by step", and his belief was in the importance of the thinking process.

3. SCIENCE IS THE CONCEPTUAL RE-ORDERING OF NATURE

Carl had taken a course on philosophy of science when he was an undergraduate, and spoke of his ideas about science and science teaching in very articulated ways, substantiating them with many examples, as revealed in the following excerpts of interviews.

> The whole business of science is the conceptual re-ordering of nature… Using figures and symbols to simplify the relatively abstract things and to make them more concrete. This is an ever-lasting theme of science… Learning the conventions of the scientific community is a part of science… Learning to use the Punnett square is not just learning a convention but also that biology has made use of some statistical tools. This represents a major breakthrough in the study of biology – the use of quantitative study. This is a very interesting topic for the history of biology or philosophy of biology.
>
> In fact, every person is a scientist in a way though he may not have studied any science… The only difference that distinguishes

> serious science from less serious science is that there is theory making in serious science; some abstract theory. In addition, serious science can make use of abstract theory to predict other situations or to bring in new experiments… Science can have many different sources of origin.

4. BROADENING THE DEFINITION OF PRACTICALS

Carl had quite a unique view of what the practical aspect of science is, and the role it should play in science education, as revealed in the following excerpts from various interviews concerning instances of his practical lessons:

> In Cantonese the word practical has a number of meanings. In the broad sense, it means practical learning experience; that is, through some hands-on manipulation to learn something, including skills like dissection, etc. Second, through practicals, students can consolidate what they have learned from theories. Third, practicals can broaden their knowledge. For example, we make use of agar in this practical, but we talk little about agar in theory lessons. So I make use of the opportunity of this practical to deepen their knowledge about agar and to show them some alternative ways to learn biology. On the one hand, practicals are important for extension of knowledge. On the other hand, they stimulate awareness of the different ways to retrieve information, like this time, I use CD-ROM as a resource. In short, I am broadening the definition of practicals. I think that besides the so-called practical experience of hands-on activities, we are also talking about multimedia, simulations. Some of them may not be real practicals, but are still very good practical components.

> The experiment on papain requires us to extract enzyme from different animal and plant sources. It is amazing that the same enzyme can be obtained from such a diversified source of materials. This is a realization of what is referred to in science as the concept of unity and diversity. I was very impressed by this experiment when I was still a student, when the teacher told me that the black and white negative could be digested by enzymes.

> To me, these two things were totally unrelated. You know, I had been taking photographs for so many years but I had never imagined that the surface layer of the negative was gelatin and that it could be used for measuring enzyme activity. I don't know if it is this revelational experience that has affected the way in which I look upon the role of practical work. So, I have been trying to put together things that appear to be unrelated according to our daily life experiences, but in fact share some common characteristics; and that those common characteristics are united through biological knowledge. This is a very interesting realization. Of course, good quality practical work should also be able to stimulate a student to think of related aspects after carrying it out; that is, to motivate them to think further.

5. CHARACTERISTIC FEATURES OF PRACTICAL WORK TEACHING

The amount of Carl's interaction with his students during the TAS practicals was the third highest of the teachers in this study. After Alan and Bob, he touched least on assessment-related issues during the TAS practicals. Of the T-S interactions related to practical work teaching, 45% was focused on concepts underlying scientific research, 29% on the procedural aspects of the practical, 19% on biology theories underlying the practical and about 2% on scientific attitudes. Most of the time, Carl was guiding students to the answer (61%), rather than telling them the answer directly (10%).

Carl strongly emphasized developing in students an understanding of the concepts underlying scientific research and consolidating the biology theories taught. He also emphasized developing in students the basic skills necessary for scientific investigations, as well as inculcating in students appropriate scientific attitudes. He also wanted to broaden the knowledge base of students through practicals.

5.1 Enhancing understanding of the concepts underlying scientific research

One prominent feature of Carl's practical work instruction was the very lengthy (often more than 30 minutes) pre-lab and post-lab discussions he had with his class in the practical sessions proper. In fact, Carl was the only teacher in this study who held post-lab discussions with students on the results of their experiments before asking them to write up their reports individually. During these discussions, Carl placed a lot of emphasis on developing in students an understanding of the concepts underlying scientific research, as revealed in the following representative episode.

Episode 6-1 – It is scientific method that is important

This was a problem-solving practical in which students had to rank order the eight unknown sucrose solutions in an ascending order according to their concentrations as determined using the potato tissues given. The pre-lab discussion began with Carl addressing the class in the following manner:

TWC: Now, like other practicals we previously had, we will spend the first period or so discussing how to do the experiment... There will be no handouts for you. I hope through discussion, I can guide you through the experiment... Even though the experiment is easy... it is scientific method that is important. It is the idea that leads to the experimental design that is important...

After this introduction, Carl asked a student to suggest a method which he then used as a springboard for the whole class discussion, as follows:

TWC: I need comments about her method. Is it workable? Mandy, what do you think? Is it workable?

S1: Yes.

TWC: You think it is very good already. Can you improve on that? It is okay but obviously it is not perfect. What are the shortcomings? What are the limitations of her method? It is okay in general but.

S1: [...]

TWC: What did she say? She said, "You put a potato strip into a solution..." Can you improve on this? Any suggestion?

S2: Put 2 or 3 strips into the solution.

TWC: She says one, you say 2 to 3. Why not 4 and 5 and why not 5 and 6? (Class laugh)

S2: Five is too many and the beaker is too small.

TWC: The beaker is very small (laughing). Well, 2 to 3. That is an improvement, why? The question is: What is so good about having more potato strips than having only a single one. Obviously, it is good. But why is it good?

S3: It is more accurate.

TWC: Yes, it is more accurate. But why is it more accurate? (No response from the class)

TWC: Why is it accurate? Even though we are running out of time, I want you at this point, to spend about a minute discussing with your classmates why using a greater number of potato strips is more accurate? Why is it good? Okay, I will come back to you after one minute. (Group discussion)

TWC: Any good suggestions? Of course, it is more accurate, but why? (No response from the class)

S4: More strips (?)

TWC: Yeah! More strips you can take the mean. The mean of what?

S4: The mean of the (?)

TWC: Are you suggesting that suppose you take 3 strips, you measure the weight change of the first strip, and the second strip, and the third strip. You have 3 figures and you take the mean of them?

S4: Yes.

TWC: Well, well, I can't say that it is wrong. It is obviously correct, quite (laughing). Any other suggestion?
(No response from the class)

TWC: Now, look at me. On my left hand, you have a single strip. On my right hand, I have 3 strips. Okay. You put them into the solutions and measure the changes in weight. Which one will give you a larger change in weight?

S5: The one with 3 strips.

TWC: Yes, if you have 3 strips then the total weight change will be more. And that would increase your accuracy level. Because the weight change of a single individual strip may not be that much. A single strip is how many grams? You can imagine, right? With a bigger mass, you have a bigger change in weight. That would bring us to another question.

TWC: How many strips will be desirable? Is it 3? Is it 4? Is it 5? If you say 5, then why don't you say 10? (laugh) If you say 10, then why don't you say 20? If you say 20, and we can go up forever, right?

TWC: Obviously, the higher the initial mass, the greater will be the change in weight. But we are limited by our apparatus. We will be using these little beakers. In fact, we can use bigger beakers. But we don't have enough of those. We can only afford these small containers. In each of these, I would suggest maybe not less than 3, maybe at least 5. Because if you have too many, you simply can't put them in. Maybe 5 or 6.

TWC: But it is most important to, ... , if you have 1,2,3,4,5 and 6 so on, the number of potato strips in each beaker must be the same. Is this essential? So, you have 5 here, you have 5 here.... Is this really essential?
(No response from class)

TWC: I want you to raise up your hands to show your approval or disapproval of this. The statement is: You need to put exactly the same number of potato strips into each of the immersion solutions. Is this essential? Those who say yes, hands up.
(Some pupils raise their hands.)

TWC: Those who say no.
(Some pupils raise their hands.)

TWC: Some say yes, some say no. So, you. (The teacher is pointing to a student.) Tell me why not? Why not?

S6: Because we only take the change in weight in the potato strips.

TWC: Very good answer. You are more or less at that.

TWC: We take the change in weight in each setup. Even though the initial weights among the setup are different, as long as you have the percentage changes in weight, it is okay.

TWC: However, for obvious reasons, you will not purposely vary them so much because a comparable starting point is always a good point for minimal comparison, right? So, we'll try to put every setup with the same amount of potato tissue at the beginning. What I suggest is that even if the initial mass of each treatment deviates a little bit from one another, it is still okay as long as you calculate the initial mass, the final mass and you take the percentage changes in weight.

TWC: That is a very important consideration. So, those who suggest that it is absolutely essential, have to rethink about that.

In a similar vein, Carl often rounded off his pre-lab discussion in the following manner:

TWC: Any questions before you start? And we will work through some of the logical thinking of the experiment during the waiting time.

Carl's emphasis on "discussing with students" can be revealed in the following episode, where he was sort of complaining why the students were working so slowly.

Episode 6-2 – If a discussion is not required, what's the purpose of doing an experiment?

T: Is it the case that you don't have to stay after school in order to be able to finish writing up your report in the chemistry practical [like we do in biology]?

S: Yes.

T: Before you leave? I mean the full report.

S: No. It is unlike the biology report. A lot of information is there already.

T: You mean there is a handout already. You just need to enter the figures there. But is there any discussion?

S: Not so many things as what are required by biology. No introduction, etc.

T: But is a discussion required?

S: No, it is not required.

T: Not required!? If a discussion is not required, what's the purpose of doing an experiment then?

The following episode was typical of a post-lab discussion between Carl and his class where he would try to point out to students the general principles and approaches in interpreting and presenting the results, instead of actually interpreting the results for individual students.

Episode 6-3 – This is corroboration

TWC: Now, I want all of you to take a look at the eight beakers. What did you discover?

TWC: There is a visual observation. What did you discover? Chung Him.

S: Some potato strips float and some sink.

TWC: Some float and some sink. Okay, what does it mean when the potato strips float?

S: The solution is more concentrated than the others.

TWC: A more concentrated sugar solution, why? What is its density?

S: It has a density higher than those with a lower concentration.

TWC: Higher density, yes. This is corroboration. It tells you that there may be something wrong if this set of results does not tally with the other set of results that you got using the concept of water potential. Okay?

At a later stage during the discussion, Carl returned to the same point again:

TWC: Furthermore, in your result section, in fact you could have put down what I called observable comments. Remember, previously I have told you that in some of the setups, the potato strips floated and in other cases, they sank. In your table, you can put that in as well. In your discussion, if you can relate your results to observable data, it will lend more support to your conclusion. Sometimes, even if you get some errors, maybe, in your weighing, with your observable results, you can still cross compare between them to get a full picture [of where it has gone wrong].

5.2 Facilitating understanding of the biology theory behind the practical work

Besides emphasizing developing in students an understanding of scientific method and its underlying thinking, Carl also placed emphasis on consolidating what the students had learnt about the relevant biology theories through discussion. The following is one of the representative episodes.

Episode 6-4 – What are the possible reasons…?

TWC: Simon, what are the possible reasons for the differences in amylase activity in different parts of the gut?

S1: May be certain segments of the intestine can secrete amylase, other segments do not. Or, it is a difference in the amount being produced.

After Carl had elaborated a bit on the relevant biological theory, the discussion moved on:

TWC: Peter, what do you think? What are the other factors that could explain the differences in activity?

S2: It may be due to the different rates of production of the enzyme. Some parts produce more amylase. Other parts produce less.

TWC: This is one explanation. What are the other possible explanations? Mandy?

S3: (?)

TWC: What do you mean? You mean at different parts of the intestine you have a different pH?
(S3 nods her head.)

TWC: Is that true? Maybe. How can you prove if your idea is correct or not?
(No response from the class)

TWC: Now, if your explanation is true then it is very easy to prove that. You simply dip a piece of pH paper into the extract to test the pH.

TWC: But you have carried out this step, right? So, you can write this down in your report saying that this is a plausible and testable hypothesis. This is a testable hypothesis because in the gut of the vertebrates, there are differences in pH in the different parts. So, you can test to find out if this is the same for the invertebrate gut?

TWC: In fact, I don't know. In fact, I don't know either. But you can test this out very easily, right?

TWC: Now, why can difference in pH be a possible explanation for the differences in the activity of the amylase in the different parts of the gut?

S4: The gut has some chemicals secreted to denature the enzyme in the lower part of the gut.

The discussion moved on for a few more minutes, concentrated on the underlying biological theory.

5.3 Developing basic scientific skills

Carl also emphasized developing in students the basic skills necessary for scientific investigations. These included not only manipulative skills but also skills in reporting and the like, as revealed in the following excerpts from various lessons.

TWC: Since this is your first time running a chromatogram, perhaps after you have started running it, don't be bothered by other things. Just observe how the solvent and the pigments move up. What is the rate of upward movement? Does it move up straight or not? When does the pigment spot begin to separate into different bands? That is, note down all your observations. … Note down all these points. There are all the original observations of your own experiment.

TWC: It seems that you people do not have a scientific habit. Mark with a pencil on your chromatogram your name, at least, to avoid confusion.

TWC: A good report of the results should tell the maximum available information from your agar plate. Obviously, you will have information about the size of the clear spot if there is any, right?... So you will have 6 columns in your table... Of course, there are other ways of reporting the results. Like you can make a drawing on graph paper... You should report all the observations... with some explanations... A table, a diagram and, of course, a photo will make it perfect!

TWC: Just now, when I assessed your raw data tables, I found that half of the class had done it very badly... This made me rather angry... This is not the first time you have done this type of experiment... You are not making dishes for your dinner, you are carrying out an experiment! … I should have demanded more from you people even though you may feel a bit more nervous.

5.4 Developing scientific attitudes

Carl also paid great attention inculcating in students appropriate scientific attitudes, in particular, an objective attitude towards the experimental results, as revealed in the following episode.

Episode 6-5 – Offer an explanation for the unexpected results
This was a practical on separation of leaf pigments extracted from plants using the paper chromatography method:

S: Sir, the colour pigments are not moving vertically upward [in the chromatogram, as expected].

T: This may not be a bad thing. There isn't in this world a case of no failure. Stick all the failure cases in your report and comment on them individually. Why is it like that? … Examiners would like to see how you could explain those failure cases and your suggestions on how to avoid those failures...

Another student approached Carl and asked:

S: Sir, why is it that there is a separate yellow pigment that's gone up here?

T: Well, you can put forward a hypothesis. Say, if this is not due to your experimental errors, then it could mean that there is a sub-type of carotene even for the yellow carotene.

S: Are you kidding?

T: But this is what you observed here. You can go and see if others are obtaining the same results or not. If that is the case, this can be a plausible explanation. And it could be a hypothesis. Right?

At a later stage, Carl made the following announcement to the class:

TWC: Never mind about the failures… Offer an explanation for the unexpected results. I don't have any comments here. I'd rather leave it for you to make your own comments in your report.

5.5 Broadening students' knowledge base

A unique feature of Carl's practical work instruction was that he would make use of the periods when students were waiting for the final experimental readings for other activities that were not directly related to the aim of the investigation *per se*. For example, in a practical on investigation of protease activity in different kinds of materials, Carl made use of the waiting time for two things: (1) to show a short video on how the Japanese make use of casein (a milk protein) to make ties and clothings etc., (2) to brief students on a handout containing information about kiwi, papain, meat tenderizer and agar; and tell the students that all this information was retrieved from CD-ROMs. According to Carl, his intention was to broaden students' knowledge base about the materials that they were using in the practical.

6. ASPECTS OF THE NATURE OF SCIENCE EMPHASIZED

Carl often talked about the nature of science in the course of his discussion with the students on the practical they were performing. He placed a lot of emphasis on (1) the importance of keen observations in the process of science, (2) the importance of putting forward plausible hypotheses and bringing in new experiments to test for new ideas, that is, the on-going nature of science, (3) the repeatability nature of science, and (4) the imperfect nature of experimentation and its limitations; as revealed in the following episode.

Episode 6-6 – You have to accept that as one of the limitations of the experiment.

S: Sir, do we need to use the same potato for all the treatments?

T: In theory yes. If you could have a gigantic potato for all the setups, it would be good. But for practical reasons, you have to accept that as one of the limitations of the experiment. Right?

7. STUDENTS' ROLE IN THE PROCESS OF LEARNING

As reported above, Carl often held lengthy discussions with the whole class. Though he was the dominant figure in controlling both the foci and the pace of the discussion, he often tried his best to involve as many students as possible in the discussion, for example, asking students to take turns to express their opinions. At points where the students' responses were not forthcoming, he would suspend the whole class discussion and ask students to have small group discussion and then resume the class discussion at a later stage, as revealed in the following episode.

Episode 6-7 – I'll leave you to think about it

TWC: So, with these ideas in mind, do you know how to do this experiment? If yes, I can leave it to you.
(No response from the class)

TWC: With all these ideas, are you confident to solve the problem posed to you?
(No response from the class)

TWC: Maybe, I'll leave you to think about it for 2 to 3 minutes. You may discuss it amongst yourself. After that, maybe someone will come out and give us a brief outline of the major steps.

Though Carl might have provided students with many insights about the practical during the discussions, he was still mindful of leaving room for individual students to make their own decisions in the course of the practical, as revealed in the following excerpts from various lessons.

TWC: You think of a way to avoid the possible confusions. This will not be taught, otherwise you don't have to think at all. You think of a good way; all the available materials are on my bench. OK! 1-2-3, start!

TWC: I won't teach you how to do it. You think of a good method to present the result of the relative amylase activity such that people can understand it once they look at it. Okay?

TWC: Okay. I leave the rest for you to think about it yourself.

8. CHARACTERISTIC FEATURES OF ASSESSMENT PRACTICES

On the whole, assessment activities did not constitute a prominent feature of Carl's practice, though he did make it known to his students when he was going to assess them, as revealed in the following excerpt.

TWC: Today I will also be assessing your practical skills as well. Try to spread yourselves out so that …

But Carl did not make the assessment setting like that of an examination hall. Not only did he allow the students to look up references during the practical, but sometimes, he even encouraged his students to consult their classmates, as revealed in the following excerpts from various lessons.

TWC: Now, while you are stationed at your place, it is not a problem for you to walk around to look at your classmates' work. You can go around to find out how your classmates are tackling the problem... Don't just sit there. This is now an era where information flows in all directions. You won't learn by just sitting there...

TWC: If you are conscientious enough, you can go around and do a survey of the results of each of your classmates... You can take a frequency count of this...For example, how many of your classmates got a result with the descending order of enzyme activity from the foregut, midgut to the hindgut? And how many got a result with..., etc.? This will help you to generalize your conclusion when you are discussing your results. Right?

At times, however, Carl did find himself in a dilemma about whether to allow students to discuss among themselves or not, as revealed in one of the lessons.

TWC: In fact, I have encountered a problem here. I took this practical as an assessment. But some of you talked to each other. Actually this is difficult to avoid, so I take the middle road. I hope that you can write up the report mainly on your own. But if you want to ask one or two words or one or two points, it is difficult to stop you from doing this...

As evident here, Carl tended to adopt a tolerant stance on this matter. In a similar vein, sometimes he allowed his students to complete their reports at home instead of asking them to stay well beyond the normal school hours (3:20 p.m.) to finish the report in class, as revealed in the following excerpt.

TWC: If I ask you to complete the report here, you will have to finish by 5 p.m. or so. I think this is no good every time like this. Now, listen very carefully. Your report must arrive at my table before 8:25 tomorrow morning. I will not accept any late report... You can phone your classmates if you can't remember the discussion points. But don't plagiarize. I can immediately tell if you copy each other's work. I know all your standards.

As evident from the last two excerpts, Carl was aware of the possible problems associated with his way of carrying out the assessment, namely, maintaining fairness of the assessment. Nevertheless he had chosen to opt for a way which he thought would better facilitate his students' learning and help him to achieve his educational goals.

9. BELIEFS UNDERLYING CLASSROOM PRACTICES

9.1 Beliefs underlying teaching of practical work *per se*

As evident from what has been presented above, there were very close matches between Carl's practices and his various beliefs. In particular, the interactive way in which he guided his students during the lengthy class discussions tallied with his belief in "making use of the teaching situation to guide students to progress step by step". This also explained why he had some reservation about the practicability of implementing the investigative approach as advocated by the TAS, given the constraints imposed on the teacher; that is, the requirement to assess the students on a sufficient number of occasions, as he put it:

> The TAS emphasizes the scientific processes, but I have doubts about its effectiveness because modeling is very important to the pupils at their early stage of learning. But under the present TAS, many teachers will rapidly do away with this modeling, because it is not possible for them to carry out too many practicals. To a certain extent, the pupils may be required to think about the procedure even before they are taught the basic techniques necessary for carrying out a particular type of experiment. You may say that we have skipped one or two steps. Hence, the effect of learning may not be achieved.

He thought that this was grossly unfair to the students because "there is not a teaching phase before the assessment takes place". In addition, Carl was a bit concerned with the logistic problems associated with this approach. Hence, he had only chosen to adopt "a middle road", where he allowed "students to diverge in some practicals, but not all so that the practical would become more manageable". Carl also saw the requirement to carry out the TAS assessment as a possible constraint for the investigative approach:

> I have wanted to leave more room for students to make their own decision. However, there is a conflict. In an assessment, you require students to complete everything within a short period of time. That would give them little time to trial out what is the appropriate enzyme-substrate ratio. This is going to be a bit too difficult for them. Thus, I have to provide them with this information instead of asking them to find it out themselves.

Nonetheless, a strong relationship existed between his teaching of practical work and his beliefs about the nature of science, for example, his belief about the on-going nature of scientific development. Equally, his beliefs about the role of practical work in science education were often brought into his focal awareness. Particularly noteworthy was that he treated practical work in its broadest sense to mean "practical learning experience", through which students not only learn some skills or consolidate what they have learned from theories, but become aware of some alternative ways to learn biology through multimedia, simulations, etc.

In addition, Carl did not choose experimental materials randomly. Instead, he had been trying hard to put together things that appear to be unrelated according to daily life experiences, but which in fact share some common characteristics that are united through biological knowledge. This was related to his belief in what is referred to in science as the "concept of unity and diversity". Not least, in his priorities, was his belief that a good quality practical should be able to stimulate a student to think of the related aspects after carrying it out; in other words, to motivate the student to think further. To conclude, among other considerations, those prominent in Carl's focal awareness of how best to teach practical work were: (1) teaching students the concept of "unity and diversity", (2) "broadening students' knowledge" and (3) "motivating them to think further".

9.2 Beliefs underlying assessment practices

As mentioned earlier, assessment activities did not constitute a prominent feature of Carl's practical work instruction. This was because Carl did not see differentiating students by their performance as the prime aim of the TAS. As he put it, "It won't be a problem even if all students are performing well in a particular assessment." More importantly, Carl had taken on firmly the idea that the aim of "the TAS is to integrate assessment with teaching and learning". This accounted for the tolerant stance that he had taken regarding many of the assessment actions. Indeed, many of these actions were based on pedagogical rather than assessment considerations. For example, Carl did not like to mention marks during the practical because he thought that it was a kind of "threat" to the students and that it was of "no help at all in improving their skills, but affecting their subsequent emotional state". He allowed his students to look up references during the practicals because he believed that looking up references was "what the scientists do day in and day out" and thus was an important skill for students to acquire. On the issue of whether to allow students to discuss among themselves, Carl saw that as beneficial to both teaching and learning. This was because this could help in "easing the teacher's burden of responding to requests for help from a large class of students". Carl also thought that there was no need to avoid discussion with individual students as this was only "a sort of moral support" for the students, and that "differentiation will not be affected by one or two points of discussion with the teacher". For Carl, discussion only meant "guiding and focusing students onto the main points and not telling them the answer", and hence would not affect the TAS assessment. This explained why Carl did not mind discussion with his students. In fact, Carl was the only teacher in this study who held lengthy discussions with his class on their experimental results before they were asked to write up their reports individually. This was in line with his belief in "making use of the teaching situation to guide students to progress step by step".

10. AN OVERVIEW OF CARL'S CASE

Carl had a very articulated view about the nature of science. His ideas of science as the conceptual re-ordering of nature and of its theory making nature and its predictive capability were particularly notable. Besides training students' manipulative skills and consolidating the theories they learnt, Carl saw broadening students' knowledge base and realization of the concept of "unity and diversity" as important goals for practical work teaching as well. His beliefs about teaching and learning were mostly learning-focused. In particular, his belief in the importance of active learning on the part of the learner was in line with his vision of lifelong education. In addition, he was also committed to all-round education.

There were close matches between Carl's practices and his various beliefs. Prominent in his focal awareness of how best to teach practical work were: (1) teaching students the concept of "unity and diversity", (2) "broadening students' knowledge base" and (3) "motivating them to think further". These were all very much related to his personal interest in the philosophy of science, his vision of all-round education and of lifelong education. Also evident in Carl's practices was his belief in "making use of the teaching situation to guide students to progress step by step" and the importance of modeling to the students at their early stage of learning.

Above all, his understanding of the TAS as a pedagogical reform as well as an assessment reform was central in determining what he would bring forth to figure in his focal awareness and what would recede to the background during his teaching of practical work in the context of the TAS.

In particular, he did not see differentiating students by their performance as the prime aim of the TAS. He strongly believed that the main idea of the TAS was to integrate assessment with teaching and learning. This explained the seemingly tolerant stance that he had taken regarding many of the assessment actions. Indeed, many of these actions were based on pedagogical rather than assessment considerations.

Chapter 7

DAWN

The evolving teacher

1. BIOLOGY IS FUN... BIOLOGY IS NOT DIFFICULT

Dawn was a lively and friendly teacher in her early thirties. She interacted a lot with students, usually smiling and making gentle jokes and understated remarks of encouragement. She had a B.Sc. degree and a M.Ed. degree, and had spent her first year of teaching as a laboratory teaching consultant, which she found very useful experience for her A-level teaching in subsequent years.

When asked about her vision as a sixth-form biology teacher, Dawn answered in the following way, "...to help students get through the examination, and ensure that they do not hate this subject. It would be even better if they can cultivate an interest in it." This was in line with the answer she gave when asked to give examples of when she felt she had done a good job in science teaching.

> I have introduced to students the amazing world of life through extracurricular activities. I led a group of students to participate in an inter-school competition on environmental protection. We were one of the 20 finalist schools... It's hard work... But we made it. That gave me a sense of achievement.

Despite her rich working experience in laboratory teaching, Dawn was still interested in attending in-service training courses related to the teaching

B.H.W. Yung, (ed.), Assessment Reform in Science, 87-100.
© 2006 *Springer. Printed in the Netherlands.*

of practical work and had attended two such courses recently. This was reflected in how she described herself as a teacher.

> I am a teacher who would like to try out new ideas. I am quite open-minded. I won't resist. I am a friendly teacher. I am quite a lively teacher. I have three mottoes: Biology is fun. Biology is part of everyday life. Biology is not difficult.

2. THE ROLE OF PRACTICAL WORK

Indeed, Dawn's three mottoes matched quite well with her beliefs about the role of practical work in science teaching, as can be seen in the following excerpts:

> The most important thing in practical work is to be able to give students a sense of wonder. This can stimulate their interest in the subject… Biology without practical work is boring.

> It is very important to relate students' everyday life experience with the practical work that they are engaging in. I feel more and more strongly about this. If knowledge is put to them in an 'academic way', from a far away place, and is out of reach to them, this is not conducive to students' learning…

> It is also important to allow students to carry out practical work according to their own design because it is a matter of ownership. It might be a bit difficult for them but it is fun. They have to learn from mistakes.

Besides developing students' interest in the subject, Dawn saw practical work as contributing to "making things concrete and visible to students and by integrating it with the teaching of theory".

3. SCIENCE IS DISCOVERY

To Dawn, science consists of two major components, "One is the object of study. The other is the way to deal with it; that is, the methodology." She emphasized the "scientific method" in her teaching, which was related to how she viewed the importance of science learning compared with other types of learning.

> Science learning is a learning for the whole life. This is a kind of method which you can make use of to master other kinds of learning. You don't necessarily need to carry out any experimentation. You can carry out observations. You know how to process the information in an objective, reliable way. You can make generalizations.

And she described her role as a science teacher in the following way:

> Students always ask me the question 'why'. I tell them that I don't know. I really don't know. I stress to them very much that science is concerned with discovery... Starting from Secondary One [i.e. Year 7] onward, I emphasize to them: What is the limitation of science? What is the method that we are using to learn it? I emphasize that these are things that I want them to learn besides the knowledge in books.

4. LEARNING IS THE DUTY OF THE STUDENTS

To Dawn, teaching was very much an interactive process between the teacher and the students, in which the students had to play an active role, with the teacher serving as a facilitator. In her words, "Learning is the duty of the students. You have to force them to think and not to stuff them with facts; to create in them an inner desire, a passion to learn." Dawn also thought that interactions amongst students themselves were equally important in their learning. She valued peer learning pedagogy so much that she regarded her interactions with students as a sort of peer coaching, in which she herself learnt also in the process. Though Dawn strongly emphasized the active role of a learner in the learning process, she did not in any sense try to downplay the role of the teacher in the process.

> Teaching plays a central role in learning… The teacher's role is to induce students to think, to facilitate their thinking. There should be an interaction. The interaction in a sense is the teaching. It is also facilitation… Teaching without learning is nothing… I cannot just put all the blame on the students… We have to be their mothers, we have to be policemen, we have to be their friends, etc.

5. CHARACTERISTIC FEATURES OF PRACTICAL WORK TEACHING

Dawn emphasized inculcating in her students the appropriate scientific attitudes as well as enhancing their understanding of the concepts underlying scientific research. She also grasped every opportunity to consolidate students' understanding of the biology theory underpinning the practical work. The amount of Dawn's interactions with her students during the TAS practicals was the fourth highest out of the teachers in this study. She was also fourth out of all teachers in relation to the proportion of T-S interaction concerned with assessment-related issues. Of the T-S interaction that was related to the teaching of practical work, 7% was spent on the procedural aspects of the practical, 33% was focused on concepts underlying scientific research, 14% on biology theories underlying the practical and about 7% on scientific attitudes. Most of the time, she was guiding students to the answer (73%), rather than telling them the answer directly (7%).

5.1 Developing scientific attitudes

Dawn gave great importance to developing in her students appropriate scientific attitudes. In particular, she spent quite some effort in persuading her students to treat experimental results in an objective manner, and not to attribute unexpected results easily to experimental errors, as revealed in the following representative episodes:

Episode 7-1 – I am not asking you to make up a story but to make deductions

Dawn was discussing with two students how to interpret the results of a practical on food tests:

S1: They [the amount of ppt.] look the same in these two tubes.
S2: (?)
T: So, you two think yourselves about what you should include in your discussion.
S1: That means this increase is due to the increase in sucrose content?
T: Yes, make your deduction confidently. If your deductions are logical and reasonable, I will certainly give you marks for that.
S1: Then, how should I make up the story.
T: No, I am not asking you to make up a story but to make deductions.

Episode 7-2 – There is no need for you to worry too much about the results

S: Miss, can we work in groups? I mean each of us do our own experiment. But at the end, we choose the best results amongst us for writing up the report.

TWC: No, I don't think so. There is no need for you to worry too much about the results. It doesn't matter to me if there are good results or bad results. This is because there must be results.... It is of course good if you can get the expected results. It is not a problem at all if you don't get the expected results. Right? I have been always stressing to you that if your experiment doesn't work, you have to review and comment on what are the possible errors. If you can't figure out what exactly the errors are, then maybe you have to repeat the experiment again. Understand?

TWC: ... Of course, we have some expected results which are reasonable guesses. They are good guesses. But there are many variables in the world. This is the wonder of the living world. That is, you just don't know what is inside the living organism. You just don't know before you try it. So, don't worry too much. So, don't give me the good results of your classmates. Let me have a look at your results even though they may not be good results.

The following episode is particularly reflective of Dawn's effort in this area. It started off with a student getting an unexpected result for the control setup, where she wouldn't expect to have any enzymatic activities at all.

Episode 7-3 – Report this faithfully in your report

S: Miss, this is the control setup with water. But it foams [which is unexpected].

T: Really? Have you added something wrong into it?

S: No, I didn't.

T: It may have been contaminated then.

S: But it is really water.

T: Then report this faithfully in your report or repeat the experiment with some other water. You try to test for the water. This is extra unusual water. It is specially made!

The student followed Dawn's advice and repeated the setup for water but then got the same result:

S: Still, there are many bubbles rising up. What should I do then?

T: This is very problematic then. Very problematic. What is this?

S: Water.

T: Water! How come that water will give you so many bubbles? Did you add other things into it?

T: What kind of water is it? I really can't help. You have to take the results if you still get the same results after you repeat the experiment. Take it as it is, though I think that this is quite unusual.

T: Now, since you have time, maybe you can repeat the experiment using water from another group. I think this water is rather unusual. [...] If it still gives you the same result, just report what you get. Tell me that you encountered some strange water.

The student repeated the experiment again but using water from the other group this time:

T: How is it this time? Is this water the 'normal' water? (laugh) Now, you can see how normal it is.

S: Yes, I have to pour away this bottle of water then.

T: No, save it as evidence. (laugh) I was just kidding.

T: Yes, it is very strange indeed. Could it be that the wash bottle is contaminated? How about if you pour away the water inside and wash the bottle thoroughly?

Indeed, Dawn was the teacher who discussed this area with students most often amongst the ten teachers in this study.

5.2 Enhancing understanding of the concepts underlying scientific research

Dawn often discussed with students the concepts underlying scientific research. The following is representative of her effort in this respect. She was discussing with a student his proposed experimental design of how to compare the sugar content in ripe and unripe bananas. It was towards the end of the dialogue which was preceded by more than ten dialogical exchanges between Dawn and the student.

Episode 7-4 – I just want to make you think for yourself

T: Is this step necessary?

S: I think this would make things more or less the same [for a fair comparison].

T: Yes, you are right? I just want to make you think for yourself why you have planned to do the experiment in this way.

S: I just can't believe that those chemistry practicals are much simpler than the biology practicals.

T: It's good then. The Biology [TAS] makes you think. The Chemistry [TAS] makes you work.

Numerous other examples were found in other laboratory sessions:

TWC: Keep track of your two hypotheses. See if your experimental data can support your hypotheses, or falsify them. Your discussion should show me how you can make use of your data to verify or falsify your hypotheses…

TWC: The fundamental question is – Why do you want to do it this way? … You have to be clear what kind of results are required to answer the aim of the experiment.

TWC: How can you make a fair comparison?… What variables have to be controlled?

TWC: You have to ask yourself, what degree of accuracy are you looking for?

T: Right, exactly. We assume that… So, you have to understand that there are different assumptions behind the different methods…

T: … So, this is just give and take. Just decide for yourself. You have to give away something when you take something. You think about it yourself.

5.3 Facilitating understanding of the biology theory behind the practical work

Dawn also emphasized developing students' understanding of the biology theory behind their observations, as revealed in the following episode.

Episode 7-5 – The "chain reaction"

It was a practical investigating the effect of different chemicals on membrane permeability of beetroot tissues. Dawn was initiating a discussion with a student while she was observing her doing her work:

T: Hey, look at this. It is very interesting. What is this tube? Is it chloroform? Did you notice something special here?

S: Yes, I did. You mean the red droplets on the surface of the solution?

T: Yes, those on the top.

S: Yes, I have tried to shake the tube to disperse them but they floated on the top again soon.

T: That means these red droplets are immiscible with chloroform. What do you think these droplets are?

S: Water.

T: Yes, where does the water come from?

S: Maybe this is the water that remained behind in the tube after rinsing.

T: So much water left behind!?

S: Would it be from the damaged cells?

T: Yes, it is possible. Why then, does it become red in colour?

S: These are solutions from the cell saps (with the red pigments dissolved in it) of the damaged cells.

T: How come the cells are damaged?

S: Damaged by the chloroform.

T: Right, that is a possible reason. These droplets cannot diffuse out now. Maybe they can later on, when immersed in water. Good.

Dawn then pointed out the above "interesting phenomenon" to the rest of the students one after another. Instead of guiding them to the answer, as she did in the above case, Dawn directed each student to the previous student with whom she had discussed the problem as the resource person in the following manner:

T: Go and ask Dorothy for the explanation. This is to test if she really understands or not.

So, the discussion about this "interesting phenomenon" spread to the rest of the class like a "chain reaction" – a term used by Dawn in her response to my probing of this episode.

6. ASPECTS OF THE NATURE OF SCIENCE EMPHASIZED

Dawn often talked about the nature of science, for example, the imperfect nature of experimentation, though in an implicit manner. Her emphasis of the imperfect and on-going nature of experimentation can be revealed in the following episode, where she was briefing students on their weaknesses in the discussion part of their reports on comparing the changes in starch and reducing sugar contents in ripe and unripe bananas.

Episode 7-6 – Distinguish between limitations and sources of error

TWC: Some of you cannot distinguish between limitations and sources of error. Limitations refer to… These are not due to your carelessness, etc. For example, how can we be sure that the increase in sugar content would really take place in a single banana… Bear in mind that we are using two different bananas – an unripe one and a ripe one. The unripe banana can contain more reducing sugars than the ripe banana at the outset because they are two different individuals. We are not actually observing the changes of a single banana during its ripening process. Right? This is a very great limitation of the experiment... Besides, we were just looking at its starch content and sugar content. We hypothesized that the sweet taste was due to the change of starch into sugar. Could there be other alternative explanations? Could it be changed from fats? Could it be changed from other food substances? We didn't perform any investigation on these? We don't know. If we really want to be sure, maybe we have to conduct other food tests too... Not until then can we have a clearer picture on the changes in the amounts of the different food substances. Do you understand?

7. STUDENTS' ROLE IN THE PROCESS OF LEARNING

In line with her belief in the "importance of allowing students to carry out the practical work according to their own design", Dawn got her students actively involved in every stage of the practical work. She distributed the lab manual to students a few days in advance and asked them to write up a proposal for their experimental design. The proposals were then marked and returned to the students on the day of the practical, when Dawn would ask the students to present their different ideas to the class. After that, students were allowed to carry out their practical work, using a method of their own choice, as revealed in the following episode.

Episode 7-7 – Four different methods have been proposed

TWC: Now, four different methods have been proposed. I suggest that you work according to your proposals. This will enable us to compare and to find out which method is a better one.

TWC: Carmen and Cecilia suggested that we should weigh the precipitate... This is a workable method [though not the mainstream one]... I have prepared the filter paper for you here... The rest of you can also change your mind now if you want to follow their method...

TWC: Okay. You decide for yourself. I think that it doesn't matter which method you are going to use. As I always say, "Never mind, just go ahead. I will always allow you to do whatever you want as long as you know what you are doing and what you are trying to achieve, and as long as you are still moving towards the same goal. Okay?

Dawn also encouraged her students to discuss among themselves, to share and to learn from each other, as revealed in the "chain reaction" episode 7-5 above and the following excerpt.

T: Yes, go and consult them. You see, they are progressing very well... They have also gone through the experience of failure. Ask them to share with you what they have learnt from their failures.

All in all, Dawn wanted her students to be active participants in the process of learning.

8. CHARACTERISTIC FEATURES OF ASSESSMENT PRACTICES

Assessment activities did not constitute a prominent feature of Dawn's practice. Besides distributing the lab manual to the students in advance, Dawn also allowed her students to look up references during the practical if they so wished. Discussions among students were not stopped and, in some cases, were encouraged, as described in the "chain reaction" episode above. In addition, Dawn often initiated discussions with individual students and helped them to fix their problems without hesitation. Though of very rare occurrence, the following are two examples where Dawn was concerned with the issue of fairness.

TWC: I hope that the experiment today is going to be a short one; and that you can all finish by 3:50 and leave on time. At most, I don't want it to be later than 4:30 p.m. This is to be fair, and so I don't want to extend the deadline for handing in the report for too long. Okay?

TWC: Now, each of you have to work on your own, though I don't mind if you discuss among yourselves in the course of your experiment. Okay? Should we start now? Okay. Start working now.

9. BELIEFS UNDERLYING CLASSROOM PRACTICES

9.1 Beliefs underlying teaching of practical work *per se*

As evident from the episodes and the portrait presented above, there were close matches between Dawn's practices and her espoused beliefs. In particular, there was a close correspondence between her practices and her belief in "the importance of allowing students to carry out the practical work according to their own design". As she put it, "It might be a bit difficult for them but is fun." However, such a realization did not come with the introduction of the TAS. At that stage, Dawn was sceptical of the plausibility of implementing the investigative approach as advocated by the TAS. This is reflected by her using the word "terrible" to describe her experience in the first year of the TAS (see Chapter 14). For the same reason, it is not surprising to find her using the word "forced" in the following interview excerpt.

> Now, we are forced to provide this type of practical for students. This is good in fact. In the past, no one forced us… But now, because of the TAS requirement, after training students on certain skills, you have to let go and ask them to do their own work… to play a more active role…

Later, Dawn changed her ideas after trying out some teaching resources that were specially developed for achieving the goals advocated by the TAS, and noticing significant changes in students' learning.

> Now I realize that it's good if there can be room for students to develop variations amongst themselves. I mean, they can think of different methods to solve the same problem. This is the variation that I treasure… I share the same view [as that of the TAS] that, in the long run, students should acquire this ability... Students enjoy doing it this way. Their feedback drives me to think about this further… I believe if students enjoy the class, they will then learn well.

> It was completely out of my expectations. I thought it was going to be a very straightforward experiment. But to my surprise, students came up with a lot of different ideas and questions which I had not thought of myself. It was the students who triggered me to think about these questions. I then guided them to think along their own lines of thought. Eventually four different methods emerged. Finally, through their own experiments, they found that the simplest method was good enough to serve the purpose... It was their own finding which they arrived at after their own investigation. I think this is really good, especially when it occurred in an assessment occasion. It did not hinder my teaching… That is the spirit of the TAS – learning through the practical work… Thus, I do not distinguish what I should do in assessment situations and what I should not do in assessment situations.

It is the last two sentences that distinguish Dawn from those teachers who still saw assessment in the more traditional way. That is, in order to maintain fairness of the assessment, there must be no interaction between the teacher and the students, or among the students themselves, and every condition of the assessment setting must be standardized, etc.

Dawn's experiences lend support to research which suggests that students' feedback can serve as a powerful incentive for promoting instructional change (McLaughlin and Talbert, 1993; Spillane and Jennings, 1997). Indeed, seeing students' interest in, and success with, the new teaching material and teaching approach as advocated by the TAS, Dawn was motivated to proceed with her effort in reconstructing her teaching. She began to re-write the lab manuals so that "students have to look at each of the steps and think of the reason why they have to carry out this step in relation to the aim of the experiment". In fact, Dawn's emphasis on "to make students think" was in line with her vision of education, as she put it:

> When I say, "The Biology [TAS] makes you think. The Chemistry [TAS] makes you work." I am referring to "Biology - the way in which I am teaching." I want to make students think. Thinking – it is how I look at education.

In short, it was her vision of education and her students' feedback that had driven Dawn to teach practical work using the investigative approach.

On the other hand, Dawn's effort to persuade students to treat the experimental results in an objective manner tallied with her belief in the "objective and reliable" nature of the scientific method as well as the limitations underlying experimentation. It is this last factor that explains why Dawn emphasized asking students to think more during the practical. This was also in line with her view about teaching and learning. According to Dawn, "Learning is the duty of the student. You have to force them to think and not to stuff them with fact; to create in themselves an inner desire, a passion to learn."

9.2 Beliefs underlying assessment practices

Much of Dawn's assessment practice can be explained by her beliefs about teaching and learning as well as the beliefs embodied in the following interview excerpt.

> I don't want the students to work under such stress…I don't want those terrible things in the first year of the TAS to happen again. At that time, I was not familiar with the scheme. Students and I

> were putting pressure on each other. They were neurotic and so was I. That is not a good way to learn…

Indeed, learning under stress was contradictory to one of the three mottoes that were guiding her teaching; that is "Biology is fun". Having gone through the "terrible" experiences in the first year of the TAS, Dawn had come to realize the need to find ways to relieve students' pressures, to protect them from anxiety and from a feeling of failure. For example, believing that students' obsessive concern with examination marks was the origin of the pressure associated with the TAS, Dawn tended to carry out assessment in a way that is unnoticed by the students, as far as possible, so as "not to scare them" and to "avoid mentioning marks during the practical".

> I don't want to use marks to threaten them... I think it is something to do with your attitude. Whether you really want to help the student, or you want to tell the student that he has committed a mistake and that you are going to punish him…I would remind them of the mistake but in the form of a joke.

Dawn allowed students to look up references during the practical because she regarded looking up references as "an important skill" that students should acquire and practice, as scientists do. In a similar vein, discussion among students was seen by Dawn as beneficial to "building up students' self-esteem and the subtle relationship among students themselves". This was also in line with her belief in the value of peer pedagogy in bringing about learning.

One prominent feature of Dawn's assessment practice was that she did not avoid giving assistance to students if she deemed it necessary. She upheld her belief that the teacher's role was to assist students' learning and that the assessment requirements should only be of secondary consideration in the process. The following interview excerpt reveals her underlying beliefs.

> If I don't do this, how are they going to learn? You know. I can deduct marks from them depending on the amount of help offered… Or, if I find all of them don't know how to proceed, I can just delete that particular assessment criterion from the assessment checklist.

It is this response that distinguishes Dawn from those teachers who tended to avoid giving assistance to students. That is, Dawn herself was 'in control' of the teaching-learning situation inside her classroom whereas, in other cases, the teachers felt very much bound by the TAS regulations, and in a sense, their classroom actions were 'controlled' by the TAS. For the same reason, Dawn wouldn't avoid discussion with individual students, as she put it:

> I practise this very often. And I like to sit next to them to discuss with them…I feel comfortable with this as long as I haven't revealed the answer directly to them. I always say, "That's what you say, not my opinion."

All in all, Dawn believed that students' learning should always be given priority over accomplishing the assessment requirements, as there are many ways of satisfying the assessment requirements. As she put it:

> In the past, I emphasized asking all students to follow a standard method so that I could grade them more easily [using a common checklist]. Now, I allow them to work with their own methods. Gradually, I have come to realize that there are bound to be variations in their methods. It's okay as long as I can make adjustments in assessing it. There can be different kinds of possible adjustment there.

10. AN OVERVIEW OF DAWN'S CASE

Dawn had a fairly well articulated view about the nature of science. She thought that science is discovery and is objective. Its method has its own limitations. She regarded science learning highly: "a learning for the whole life" because it is "a kind of method which [one] can make use of to master other kinds of learning." She had a strong view of the role of practical work in school science education – stimulating students' interest in the subject, consolidating theory learning, learning the scientific method and acquiring a scientific attitude. Her beliefs about teaching and learning were mostly learning-focused. Most notable was her idea that "teaching without learning is nothing". Her prime goal in teaching was to "help students to get through the examination" because she regarded it a bonus if "students can cultivate an interest in biology". This latter goal was entrenched in her three mottoes, in particular, her emphasis on the element of fun in the process of learning.

There were close matches between Dawn's practice in the teaching of practical work and her beliefs about science, teaching and learning. For example, Dawn's belief about science as an objective and reliable way of processing information and making generalizations was evident in her teaching of practical work. Also implicit in her teaching was her belief about the imperfect and on-going nature of experimentation. Prominent in her focal awareness were also her learning-focused beliefs. In particular, she wanted her students to be active participants in the process of learning. She believed in peer pedagogy and she asserted that learning under stress was not a good way of learning. All these beliefs were very much related to her understanding of the TAS as a pedagogical reform as well as an assessment reform. In her words, the spirit of the TAS was – learning through the practical work. Thus, she didn't avoid giving assistance to students if she deemed it necessary. She upheld her belief that the teacher's role was to assist students' learning and that the assessment requirements should only be of secondary consideration in the process. She thought that there could be different ways of satisfying the assessment requirements. All in all, Dawn herself was 'in control' of the teaching-learning situation inside her classroom, unlike those teachers who felt bound by the TAS regulations. Nevertheless, this realization of the flexibility provided by the TAS did not come with the introduction of the TAS. Rather, it was a result of trying out and seeing students' interest in, and success with, the new teaching material and new teaching approach advocated by the TAS.

Chapter 8

EDDY

The money-hunter

1. I WAS A MONEY-HUNTER

Eddy was a very practical, down to earth teacher in his early forties. Immediately after receiving his B.Sc. degree and his Diploma in Education, Eddy taught examination classes in a private tutoring school for seven years. He thought that this had "trained him up" because it made him "very familiar with the examination syllabuses". However, this experience did not help him to adjust to his new teaching life in a mainstream school where the students were mainly academic low achievers. Eddy recalled those days as a "horrifying" experience for him. He was so frightened that he decided to quit after he had taught only for 42 days there, even though he was not offered a position by any other school at that time. Eventually, he joined his present school; it was his ninth year there when this field study was carried out.

Eddy's characterization of himself in the past as a "money-hunter" was an example of his practical nature. He served as an auxiliary policeman "just for money" when he was still an undergraduate. The turning point for him was the June 4th incident in 1989, in Beijing. Not because it had made him lose a lot of money in the stock market, but because of the many young people who had lost their lives in Tienenman Square. And that had changed his view of the meaning of life, as he put it:

> My focus had always been on earning more money when I was teaching in the private tutoring school. I called myself a money-hunter. I taught also in an evening school. I seized every chance to

B.H.W. Yung, (ed.), Assessment Reform in Science, 101-114.
© 2006 *Springer. Printed in the Netherlands.*

> earn money. But after the June 4th incident, I asked myself if money was really that important. Are there any other ideals in one's life other than earning money? … It was impossible for me to achieve those ideals in a private tutoring school. Hence I decided to move to a subsidized school [fully funded by the government]. The desire to settle down, and to work in a stable environment was another reason. I didn't want to trap myself in that kind of environment any more. I wanted to explore another new environment.

2. I AM A PERSON WITH NO VISION

Interestingly, many years later when this study was undertaken, Eddy still described himself as "a person with no vision". He attributed this to his way of looking at the teaching profession as "very different from other professions". He believed that "people in this profession do not like their peers to work hard and to have ambitions. Otherwise, you will easily catch the others' attention. The principal and your colleagues will then look at you with a suspicious eye." Indeed, Eddy's teaching was very much affected by this kind of thinking, as revealed in the following excerpt.

> When I am teaching classes that are below average, I don't require them to think, to do experiments. They only need to study what I have taught them… There are tests after tests… I do this just for the sake of maintaining a high passing percentage. In a way, I am quite selfish. I might have buried some of the students' interests in studying science… Even though the principal says that he doesn't care too much about the public examination results. However, the examination results are laid on the table for everyone to peruse in the first staff meeting every year. He comments on each and every individual subject. This is a kind of pressure…

As indicated above, Eddy was a down to earth examination-oriented teacher. His examination-oriented attitude is further revealed in his explanation of why he thought he was more popular among the higher form students than among the lower form students. He attributed that to the higher form students' appreciation of his giving them "the rich body of knowledge, especially those of the examination techniques". In fact, Eddy's examination-oriented attitude had its root traced back to the days when he was still a student, as he put it:

> I was very examination-oriented. I just didn't bother about stuff outside the syllabus. My primary goal was to be able to master the examination... The focus was not on learning knowledge but on finding out how I would be examined.

Such an attitude was very much related to his pragmatic and practical attitude, as revealed in how he determined the career pathway for himself:

> To be frank, I'd never thought of pursuing further studies in biology and becoming a biology teacher. I studied biology in the University just because I obtained good results in this subject... Once, I enrolled in an intensive course leading to a pharmacy degree. But I dropped out in year one because the principal told me that he was going to promote me to a senior position in the school. This then changed my plans.

Eddy was very concerned with gaining his students' confidence in him as a "knowledgeable" teacher. He thought that, in order to be more efficient in teaching and learning, "the teacher must build up a strong sense of authority" otherwise the students would not "believe in you". He stated that "a teacher needs to show students that he has confidence in himself and that he is well prepared" in order to gain students' confidence.

3. I HAVE TO FEED MY STUDENTS A LOT

Eddy held a transmission view of teaching, as revealed in the following excerpt when he was describing what the ideal student should be.

> They shouldn't just rely on my feeding them with knowledge. Of course, I have to feed them. I have to feed them a lot. But they have to digest it. In addition, they have to find extra 'food' for themselves after I have fed them.

This view of teaching as a transmission, coupled with his concern for gaining confidence in himself as an authoritarian source of knowledge, also explained why he did not allow students to choose the titles for their own individual book review projects. In his words:

> I gave them titles to choose from. I didn't want them to work outside areas that are beyond my abilities.

Eddy also thought that using an inductive approach in teaching would require one to spend more time, because it was more difficult to "control" and it was just "a waste of time" using this kind of strategy in the lower forms.

4. SCIENCE IS LOGICAL AND SYSTEMATIC

Eddy defined science as "involving a process which is not by trial and error, but is logical and systematic". He believed that science "is supported by evidence which is repeatable, and testifiable". He regarded observation as the most basic thing in science. Other common methods included comparison, classification, identification, doing calculations, etc. He also believed that there could be many methods but there must be "one fastest way of arriving at the answer. One and only one." However, he added that "not all science can be verified through experiments. Some is still theory."

5. THE ROLE OF PRACTICAL WORK

Regarding the role of practical work, Eddy confessed that he didn't give detailed thought to it when making up the year plan. In his words, "It is just based on random choices." Upon further probing, he indicated that practical work was effective mainly in: (1) training students' on certain basic laboratory techniques, (2) cultivating students' scientific attitude, and (3) developing their interest in the subject. As a result of inadequate time, he thought it would be more difficult to achieve the other roles of practical work in science teaching, like developing students' conceptual understanding, verifying theories, rectifying misconceptions and practising problem solving.

6. CHARACTERISTIC FEATURES OF PRACTICAL WORK TEACHING

Eddy was rather activity-oriented in the sense that he emphasized the procedural aspects of the practical work. Rarely did he probe deeply into the conceptual side, be it related to concepts underlying scientific research or biology theories underlying the practical work. The amount of Eddy's interactions with his students during the TAS practicals was the fifth highest of the teachers in this study. He was also the fifth of the teachers in relation to the proportion of T-S interaction on assessment-related issues. Of the T-S interaction that was related to practical work teaching, 45% was spent on the procedural aspects of the practical, 31% focused on concepts underlying scientific research, 14% on biology theories underlying the practical and about 5% on scientific attitudes. Eddy guided his students to the answer of their questions in about 47% of the T-S interactions and told them the answer directly in about 16% of the T-S interactions.

6.1 Demanding attention to operational details

Quite a large proportion of interaction between Eddy and his students was focused on the procedural aspects of the practicals (45%). In particular, he was very demanding about students' performance at the level of operational details. For example, he required students to label the diagrams properly, use past tense and passive voice in writing up their reports, and so on. The following two episodes were particularly revealing with regard to the minute details he required his students to attend to in writing up their experimental reports.

Episode 8-1 – Draw a larger diagram

T: Draw a larger diagram. This is too small. It is no good. Larger diagrams are much clearer.

Episode 8-2 – It should be 0.680 and 0.330 instead of 0.68 and 0.33

TWC: You people have to pay attention to the number of decimal places that is required when you are reporting your experimental results.

TWC: Given the pipette, what is the most accurate result that you can report? How many decimal places can be reported?

TWC: Don't just do it in a very casual manner. Some of you reported 3.1 cm cube and others reported 0.33 cm cube. Don't mess around, you lot.

Eddy then circulated to check if other students were following his advice. He talked to one of the students in the following manner:

T: It should be 0.680 and 0.330 instead of 0.68 and 0.33… Do you know why? Since the total volume for the pipette is 1 ml and hence 1 small unit equals 0.1 ml, 1 small subunit equals 0.01 ml. But each small subunit is further divided into 2 units, hence each equals 0.005 ml.

6.2 Understanding concepts underlying scientific research

Though 31% of Eddy's interaction with his students was spent in this area, most of them were focused on the concept of control. In most cases, Eddy was simply reminding one student after another of the concept of control. Seldom did he probe if the students really understood the reasoning behind it. A typical interaction of this kind follows.

Episode 8-3 – What is meant by a control?

T: Timothy, why didn't you put a piece of filter paper in this set up [so as to increase the surface area in contact with the sodium hydroxide solution for absorbing the carbon dioxide released by the seeds]?

S: I think it is not necessary because it is the control.

T: What is meant by a control? All factors must be kept constant in a control.

It is evident that Eddy could have probed the student for reasons why he thought that it was not necessary to put a piece of filter paper in the control setup, as in the experimental setup, before *telling* him the concept of control. He could also have given the student some time to reflect on his question of "What is meant by a control?" before telling him the answer straight away. However, the interaction did not go beyond the level of 'reminding'.

6.3 Relating the practical to the underlying biology theory

About 14% of Eddy's interaction with his students was focused on this area. However, instead of focusing students on the key biological concepts underpinning the practical work, Eddy was mainly concerned with specific and minute, but obvious, details. As such, some students saw this kind of interaction as a disturbance rather than a learning opportunity, as revealed in the following episode:

Episode 8-4 – How come that you are so impatient?

T: Why are you removing the seed coats? What is it for?

S: For gaseous exchange, to allow the seeds to breathe.

T: Why is it no good if you don't remove the seed coat?

S: It will then hinder the process of gaseous exchange.

T: What gases are being exchanged?

S: Eyya!! The seeds have to breathe.

T: Breathe for what gas? There are many gases like nitrogen, hydrogen, oxygen and carbon dioxide, water vapour.

S: Of course, it's breathing for oxygen.

T: Then, what is the disadvantage of not removing the seed coat? Is it that the seeds cannot breathe if the seed coat is not removed?

S: Of course, not.

T: What then?

S: It will breathe at a slower rate.

T: At a slower rate.
S: Don't disturb me.
T: Huii, I just asked you a few questions and then you asked me not to disturb you. How come that you are so impatient?

As evident from the above, Eddy could have wound up the dialogue when the student correctly pointed out to him that removing the seed coat would not hinder the process of gaseous exchange (see the 4th line in the episode above). However, he did not, and continued to ask the student on very specific but obvious facts – "Breathe for what gases?" No wonder, the student was impatient with him for such a simple and unchallenging question.

6.4 Developing "scientific attitudes"

About 5.1 % of Eddy's interaction with his students was focused on this area. In particular, Eddy wanted his students to faithfully report their experimental results. However, he seemed not to have faith that his students would follow his advice. Thus, he often said to them, "Don't cheat." The following episode is particularly revealing on this aspect.

Episode 8-5 – You must be fabricating your results
It was a practical on investigating the presence of starch digesting enzyme in different regions of the alimentary canal of the grasshopper. Eddy was accusing a student, Jeremy, of fabricating the experimental results. This was because he anticipated that there shouldn't be any positive results (i.e. formation of precipitate) in the tube which the student had wrongly treated with iced water:
T: Jeremy, How is it going? Anything happens in the fore gut? How come that you got so much precipitate in the setup for the fore gut. You must be fabricating your results.
S: No, I didn't.
T: I saw you putting the tubes in iced water [and hence there shouldn't be any precipitate found in these tubes.]
S: You see. Can't you see the precipitate inside?
T: Yes, I can see that there's a lot of precipitate there. How did you add them into it?
S: How I added?
T: Huuh?! I doubt very much about your scientific honesty. You must be fabricating the results. I am sure you are.
Eddy then announced to the class:
TWC: After you have finished, leave all the nine tubes for inspection. I want to check if you are fabricating the results to fool me or not.

However, other than requiring students to be faithful to the experimental results, Eddy showed no evident concern for appropriate rigour in conducting scientific investigations, as revealed in the following episode:

Episode 8-6 – You have to consider the workload of the technician
It was a pre-lab briefing on a practical using a weight potometer to compare the rate of water loss from the upper and lower epidermis of a leaf. Eddy was suggesting to the students not to use the oil provided (to reduce evaporation of water from the open end of the potometer) in

their setup in order to save the trouble of the laboratory technician cleaning all the greasy glassware after the practical. One of the students was concerned with the accuracy of the experiment and asked:

S: Why don't you let us use the oil in our experiment since you have already prepared it?

TWC: It should be quite alright now. [Forget about the accuracy.] You see. You have to consider the workload of the lab technician too. It'd be very troublesome for her to wash the greasy apparatus afterwards. Don't just look at it from your side.

Thus, quite conflicting messages were sent to the students regarding scientific attitudes. On the one hand, he required the students to be faithful to their experimental results. On the other hand, he asked the students to disregard the issue of accuracy just for the sake of saving the trouble of the laboratory technician in cleaning the greasy glassware. What was the hidden message there?

7. ASPECTS OF THE NATURE OF SCIENCE EMPHASIZED

Eddy seldom talked about the nature of science, either explicitly or implicitly. On encountering some unexpected results, Eddy would quickly attribute them to the "sources of error", as revealed in the following excerpt. In a way, Eddy was reinforcing in his students the so-called 'right answer syndrome' instead of really encouraging them develop an inquiring mind:

T: Why do we sometimes not get the expected results? For example, I have given you three possibilities. You can write it down in your discussion. What are the sources of error? …

8. STUDENTS' ROLE IN THE PROCESS OF LEARNING

One striking feature of Eddy's practice was the authoritarian stance he adopted in his teaching, as revealed in the following episode where he was teasing a student for his alleged stupidity.

Episode 8-7 – Why are you so stupid?

T: Paul, be quick!!

S: There is a lot of solid residue. It is difficult to suck them up.

T: I did not ask you to suck up the solid residue, but the enzymes. Tell me. Where should the enzymes be? If the enzymes were to be active, they must be in the aqueous form. Why should you bother by the solid residue at all?

S: (?)

T: Yes, of course. Why are you so stupid? What should you do then?

In a similar vein, he ordered a student to stand behind his classmate to learn how to carry out a certain experimental procedure in a rather rude manner, as revealed in the following excerpt:

T: You see, how careless you are? You go and see. Your classmates are performing so well… *I am now asking you* to stand behind Alice and watch her doing this *for 10 seconds.*

Similarly, in order to facilitate his assessment, Eddy required his students to follow a standard procedure in carrying out the practicals, as revealed in the following representative excerpt:

TWC: All the apparatus and materials required for this experiment are on the teacher's bench. Don't use other apparatus which is not provided here.

All in all, Eddy expected his students to conform to what he thought was right and best for them. The students were just passive followers of the teacher, implementing what the teacher thought was most appropriate.

9. CHARACTERISTIC FEATURES OF ASSESSMENT PRACTICES

Though Eddy often initiated discussions with individual students, he did not allow the students to discuss among themselves or to look up references during the practical. The atmosphere inside Eddy's classroom was a tense one and students' concern with marks was obvious even to the most casual observer, as revealed in the following episode:

Episode 8-8 – Will marks be deducted from me if I get an extra one?

It was a practical on investigating enzyme activity in different regions of the gut in a grasshopper. The practical required students to remove the entire set of gut from the grasshopper first. But one student, Lander, failed to do so; he had about one-third of the gut damaged. Eddy noticed this and intervened in the following manner:

T: Lander. Are you having a problem?
S: No. No problem.
T: How much should I charge you [for giving you an extra grasshopper]? Are you not okay?
S: *Will marks be deducted from me if I get an extra one?*
T: What do you prefer, to carry out a better experiment or to have some marks deducted?
S: What would happen if I get an extra one?
T: Nothing will happen. All things will remain the same.
S: I want an extra one then.
T: You want it now. Be quick. Time is running out.

Eddy went away to get another grasshopper for the student. When he returned, the student asked him:

S: Will marks be deducted from me?
T: Why do you mind so much whether marks will be deducted or not?

S: It is because I think this [the damaged gut] would still work.
T: Okay, then. Just carry on with that [the damaged gut]. You have to be confident of yourself.
S: But, then you said?
T: I said nothing. I just asked you if you wanted an extra one or not.
S: What would happen if I get an extra one?
T: Nothing will happen. Why are you always mindful of whether marks will be deducted or not?
S: Just let it be now, unless I cannot carry on with my work using the existing specimen. I will get the extra one if I cannot get any result with this one.
T: Is getting a high mark the only purpose for doing experiments?
S: Yes. And I think this [the damaged gut] is going to work.

Actually, Eddy was aware of the situation himself. As he put it, "I could feel that students are even more nervous compared to their counterparts in the past when there was a practical examination [but not the TAS]. Sometimes, students did quite a lot of redundant things during the practical." Below is an example of what Eddy referred to as "redundant things" carried out by his students:

Episode 8-9 – This is to provide a safety net for me

Eddy was about to tidy up the teacher's bench and threw the left-over seeds from the experiment into the rubbish bin. A student came out to take another lot of seeds back to his workstation as some sort of fallback for re-running the experiment in case the first lot turned out not to be working properly, though he had already started writing up the report. Eddy asked him why. The student replied, "This is to provide a safety net for myself."

Eddy attributed the problem to students' obsessive concern with examination marks, "The situation now is that the students are not serious about the practical work itself but the marks that are attached to it. They are more concerned about whether or not I am going to deduct marks from them when they raise a question with me. Hence they won't ask me any question even if they don't understand." Nonetheless, the students' behaviour was not unrelated to the way that Eddy handled the assessment. Some of his assessment actions might have, in part, aggravated the situation, as revealed in the following episode.

Episode 8-10 – Do you think you are in a buffet now?

While watching his students taking the necessary materials from the teacher's bench back to their own workstations to carry out the practical work, Eddy talked to them in the following manner:

TWC: Do you think you are in a buffet now? I just don't care how much you take. Just go ahead and take as much as you want. But I will deduct marks from you for taking more materials than you need.

Though Eddy seemed to be quite concerned with the assessment issues, he was not as enthusiastic as other teachers in supervising his students' assessed work. He deliberately allowed his students to complete their reports in the library without his supervision, even if he was available. I shall return to this point in Chapter 15 to find out the underlying reasons why he did so. Overall, assessment activities constituted a prominent feature of Eddy's practice.

10. BELIEFS UNDERLYING CLASSROOM PRACTICES

10.1 Beliefs underlying teaching of practical work *per se*

As evident from above, there was a close match between Eddy's practice and his various beliefs. Prominent among them was teaching as a transmission of knowledge from the teacher to the students, and the importance for a teacher to gain students' confidence in him as an authoritarian source of knowledge.

Eddy's insistence that students attend to minute operational details when carrying out practical work, like reporting the experimental results to the third decimal place, drawing larger diagrams, labelling the glassware properly, and using past tense and passive voice in writing up their reports, tallied with his belief about the role of practical work in training students on the basic techniques. This was in line with his conception that science is systematic.

However, quite different messages were sent to the students on aspects related to scientific attitudes. On the one hand, he asked his students to report the experimental results faithfully in their reports. On the other hand, he did not allow his students to use the oil in setting up their experiment, even though he had prepared it for them already, just for the sake of saving the lab technician's trouble in cleaning up the glassware. It could be argued that he was not genuinely concerned with the goal of developing students' scientific attitudes as espoused. Instead, he was very assessment-oriented, in the sense that he would be satisfied as long as students could explain the "sources of errors" in their reports. In short, the main agenda for doing practical work was to be able to produce a decent experimental report for assessment purposes.

In relation to his espoused belief in developing students' interest in the subject through practical work, it seemed to be a rather unimportant goal for Eddy too. This was especially true for the TAS practicals where the students generally worked under a very tense atmosphere (see below).

10.2 Beliefs underlying assessment practices

Eddy believed that exerting a bit of pressure on students could make them work harder in their studies. That is why he told his students that all practicals would be assessed when actually this was not the case. He did this just for the sake of "keeping the students working hard throughout the term". This also explained why he used mark deduction as a "threat" to ensure that students did not take more materials back to their workstations than was necessary. As he put it:

> I will make use of mark deduction to deter them from doing so. In fact, I don't really deduct marks from them.

Also evident in Eddy's assessment actions was his belief in the unequal power distribution inside the classroom between the students and the teacher. This can be revealed in how he would deal with disputes over mark deduction raised by the students. He saw his role in this aspect as similar to that of the "referee of a ball game".

> Sometimes, students argue with me that they have not committed a certain mistake. But I tell them that this is it. They have to respect the decision of the referee even though the judgment may not be fair in some cases.

That is, Eddy held an authoritarian stance on this issue. This tallied with his belief in the teacher as an authoritative source of knowledge for the students, as revealed in the following excerpt where he was explaining to me why he did not allow students to discuss among themselves during the TAS practicals.

> There is nothing worthwhile for students to discuss among themselves. Also, I fear that they would copy each other's work.

Eddy's idea of "nothing worthwhile to discuss amongst students themselves" was in line with his view of teaching as a transmission of knowledge from the teacher, an authoritative source, to the students. Such a belief was also evident from the following excerpt, where Eddy was explaining why he did not hold post-lab discussions with his students on the experimental results before they were asked to write up their reports individually.

> If I discuss the results with them, it is likely that they will put down all my words, my language and my ideas.

As implied in this excerpt, Eddy took "discussion" as equivalent to the teacher telling students *all* the relevant knowledge. Once again, this was in line with his view of teaching as the transmission of knowledge. In addition, Eddy's concern about the inability to differentiate between students based on their performance was evident in his concern about students copying each other's work and producing very similar reports. Indeed, Eddy was more concerned about accomplishing the assessment requirements than facilitating students' learning, as revealed in the following excerpt.

> This time, I provided students with all materials they needed... It was something like an examination. They had to follow [the standard method in the marking scheme] as closely as possible. This is to facilitate my assessment.

Obviously, the TAS reform was perceived by Eddy primarily as an assessment reform that was not much different from the traditional practical examination. That is, its prime aim was still to differentiate students by their performance "to very minute differences". The pedagogical dimension of the reform was very much overlooked or neglected.

11. AN OVERVIEW OF EDDY'S CASE

In Eddy's own language, he is a teacher with "no vision". He was a down to earth examination-oriented teacher. He held a transmission view of teaching and regarded the teacher as the authoritarian source of knowledge. He saw observation as the most basic thing in science, which is logical, systematic and supported by evidence that is repeatable and testifiable. He had not given detailed thought to the various roles practical work might play in school science education.

There were mismatches between Eddy's practices and his espoused beliefs in the role of practical work in developing students' scientific attitudes and interest in biology. Overall, his belief in teaching as a transmission of knowledge from the teacher, an authoritarian source, to the students, was pervasive in his practices. Of paramount importance was also his understanding of the TAS as an assessment reform aimed at differentiating students' performance as precisely as possible. Thus, it was not surprising that Eddy based his practices mainly on assessment considerations rather than on considerations of students' learning. His instrumental goal of helping students to excel in the examination was in the foreground of his focal awareness, such that the main agenda for doing practical work was to produce a decent experimental report for assessment purposes. As a result, the pedagogical dimension of the TAS reform was very much neglected by him.

Chapter 9

HUGO

The examiner of a driving test

1. LEARNING IN SCHOOL IS SIMILAR TO WORKING IN SOCIETY

Hugo was a very pragmatic, utilitarian teacher in his mid thirties. He joined his present school after he graduated from university. He started teaching A-level biology in his second year of teaching. He was actively involved in the development of the TAS and had contributed to many of the in-service training courses. Hugo was in the first year of a two-year part-time M.Ed. course during the period of field study. He had chosen "School Administration" as his specialism for the M.Ed. Course. He thought that was most relevant to his recent appointment as Assistant Principal.

Hugo spoke of his vision of what education should be for his students in clear focused language and found it easy to explore educational issues. As a science teacher, Hugo emphasized requiring students "to think for themselves, to ask questions on their own initiative". But he was not satisfied with his students' performance in this respect. He attributed this to "students' low ability, lack of initiative and the way that they have been trained". He did not think that as a single individual he could do much to improve the situation unless "there are more teachers working together or the school authority can think of some practical methods to change students' way of learning". In other words, Hugo was trying to find out a solution to the problem at the system level. This was in line with his response when asked about which particular aspect of his work he would like the most help with at that point in time.

B.H.W. Yung, (ed.), Assessment Reform in Science, 115-128.
© 2006 *Springer. Printed in the Netherlands.*

> To look at science education from a broader perspective: What is the role of individual subjects? What is the vertical linkage from lower forms to higher forms? Where is the horizontal linkage of the coordination and cooperation amongst the different subjects at the same form? I think this is especially important in the lower forms.

Parallel to Hugo's tendency to explore educational problems at the system level, his classroom practices were also often influenced by his commitment to educational goals that go beyond his own subject discipline. In particular, Hugo put a lot of emphasis on "preparing students for their future working life". For example, he emphasized the necessity of training students to work under the stress of "meeting deadlines".

> When I said, "Time to finish is the first priority, otherwise forget about the accuracy." I was telling students the facts of life. There is always a time limit for us to complete a certain task. I hope that they can become accustomed to meeting deadlines. This is one of the many things that students need to be trained for in school.

For the same reason, Hugo would not warn students of the exact time left for completion of the task towards the end of the practical.

> I had made it clear to them at the beginning that I wouldn't warn them any more... This is good training for them when they work in the society. People will not usually remind you kind-heartedly in the work place. Too many reminders of this kind will not help them in the future. I hope to train them to work independently.

Hugo's perception of the work place as lacking people who would give you "kind-hearted reminders" further reinforced his belief that he should prepare his students how to adjust to the realities of life.

> When I made public the weaknesses of individual students including the marks they received, I was hoping that others would learn from the mistakes of their classmates. I have told them right at the beginning. This is just similar to practical life situation... Learning in school is similar to working in the society. They have to learn to adjust to life situation in reality.

For Hugo, one of the ways to prepare his students for their future working life was to teach them how a good citizen should behave, as revealed in the following excerpt, where he was explaining why he was very angry with a student who carried on his work beyond the time limit without his prior approval.

> This is about teaching students how a good citizen should behave. As a teacher, I have to perform this kind of extra duty from time to time… I will give extra time allowance to students to complete their work if they have strong reasons. But they must make a request first. They can't just carry out things without permission and then try to explain only after being caught on the spot. It is very dangerous for the student himself in future if he does things in such a manner. Hence I must stop this kind of behaviour.

Indeed, Hugo believed that the goal of "preparing students for their future working life" is so important that students shouldn't be solely preoccupied with their academic studies. Instead, they should participate more in extracurricular activities so as to develop other skills and qualities that would better prepare them for the future working life.

> I don't think it is good to extend the practical beyond the normal school hours… I think A-level students should not spend too much time in their studies. They should participate in extracurricular activities, serving the role of coordination, planning, contacting outside bodies and external agencies, etc. In so doing, they learn about the running of society and how to work with other people… All these, I think, will affect and benefit their future life even more… Whether one is a bright person, flexible, optimistic, positive, proactive and highly adaptive, etc., these qualities would determine whether one will succeed or not.

Hugo's shift of emphasis away from the examination results was related to his realization that "for getting a good examination result, there are in fact many factors affecting it. Gradually I realize that our effort is not as influential as I have thought before." Instead, he hoped that he could do more in other aspects: "for example, their conduct, their looking at things from a wider perspective, to cultivate in them an attitude to live a more optimistic way of life". Indeed, Hugo had "gradually moved away from biology" in seeing his role as a teacher. In his view, "when compared with knowledge in books, the value systems cultivated in students and how they evaluate a certain issue, etc. could last for a longer time. These kinds of ideas and skills can be applied in their real life situations and are thus more practical and useful to students."

2. SCIENCE IS AN OBJECTIVE METHOD TO SOLVE PROBLEMS

Hugo defined science as "an objective process or a method which can be used to solve problems... There must be certain elements like comparison, deduction, etc. but not an exact procedure... Different investigations require different methods to achieve. Simple delivery of facts is not science." Hence, for Hugo, the important goal for science teaching "is that students should know the possible ways of how to find out the answer; and how to interpret what they observe..."

3. THE ROLE OF PRACTICAL WORK

Akin to his idea of seeing his role beyond that of a "biology" teacher, Hugo downplayed the role of practical work in theory building and in developing conceptual understanding. Instead he emphasized its role on developing students' self-esteem.

> Now, I regard practical work as an independent entity which is detached from the teaching of the theory. It has become an important element that students must learn – a skill of learning how to cope with problems. Scientific method is in fact one of the methods used to solve problems. I want to make use of it as a tool for building students' interest, confidence, or giving them experience in dealing with their own problems so that they can find out where their own abilities lie. If plan well according to their abilities, this will provide them with sense of achievement; this can build up their confidence...

In fact, this was the main reason why he asked his students to carry out an individual ecology project around the campus, stating that "Giving them this kind of project is especially important for those who are not good in theory. At least, there is a greater chance for them to have a sense of achievement." Indeed, he valued practical work as a means of developing in his students a sense of achievement more than he valued its use for other purposes. This was in line with his idea of regarding practicals as a practical experience and his emphasis on "preparing students for future working life".

4. TEACHING AS TRANSMISSION OF KNOWLEDGE

In Hugo's opinion, for teaching to occur, "there must be a person, through various means, like movie, computer software or other methods... Nevertheless, the main direction should come from the teacher who guides the teaching." As a corollary, "teaching must be taking place in a teacher demonstration because during which the teacher has to ensure that students can look at the specimen and tell them which part is which; that is, there must be new information given to students." Overall, Hugo viewed teaching as a transmission of knowledge from the teacher to the learner. In other words, he was more teaching-focused than learning-focused.

5. CHARACTERISTIC FEATURES OF PRACTICAL WORK TEACHING

In line with his belief in the role of school education as preparing students for future working life, Hugo gave great importance to training students to acquire basic scientific skills as well as observing the deadline for completion of tasks. Hugo was the third, out of the participating teachers, who *least* interacted with the students during the TAS practicals. He was also the third out of all teachers regarding the extent of his concern with assessment-related issues. Of his T-S interactions, about 11% was focused on the assessment-related issues. Of the T-S interactions related to teaching of practical work, 72% was focused on the procedural aspects of the practical, 15% on concepts underlying scientific research, 7% on biology theories underlying the practical and 3% on scientific attitudes. Hugo told students the answer to their questions directly in about 40% of the T-S interactions and he guided them to the answer in about 33% of the T-S interactions.

5.1 Developing basic scientific skills

Quite a large proportion (72%) of the interaction between Hugo and his students was on the procedural side of the practicals. Most of the dialogues were related to skills and techniques that were specific to the practicals in question. Nevertheless, Hugo did emphasize training students to acquire the basic scientific skills that he considered generic in nature; that is, skills which are applicable to scientific investigations in general, as revealed in the following excerpts:

T: Fanny. Were you the last one who took chemicals from this reagent bottle? Do you see what you have done?
S: Sorry, I did not put the stopper back.
T: Next time, you better push the rubber tubing [of the Bunsen burner] all the way into the nozzle [of the gas tap]. That makes it more firmly attached to it.
S: Can we not wear the lab gown?
T: No. No. It is not very hot. It is a something you should always do when doing experiments in the laboratory, you should wear a lab gown.

5.2 Enhancing understanding of the concepts underlying scientific research

Only 15% of T-S interaction on practical work was about concepts underlying scientific research. In most cases, they involved Hugo pointing out to the students the misconceptions they had of relevant concepts revealed in their experimental procedures. Below is one representative episode.

Episode 9-1 – It's useless to set up a control 20 minutes after the experiment has started
It was a practical comparing the transpiration rates of two different kinds of leaves. Hugo noticed that a student was setting up a new experiment at quite a late stage of the lab session:
T: Stephen, are you repeating your experiment?
S: Yes, I am trying to set up the control experiment.
T: It is useless to set up a control experiment after the experimental one has been running for 20 minutes already. What's the purpose if the two set-ups are not running at the same time?
S: For comparison.
T: No. It cannot serve the purpose. You are spending time on something which is not useful. You are just wasting your own time. The two sets of data will not be comparable since they are collected at different times. Don't waste your time.

5.3 Enhancing understanding the biology theory behind the practical

Only 7% of Hugo's T-S interactions on practical work teaching touched upon the biology theories underlying the practicals. Most of these occurred during the post-lab briefings where Hugo was going over the underlying biology theories in relation to the experimental results. Usually, these briefings took place only after students had handed in all their reports and lasted only for five to ten minutes; that is, they were done in a rather hasty manner. Below is one of the representative episodes.

Episode 9-2 – Try to apply the theory to explain the result

TWC: … Besides, stomata, what else could explain the difference in transpiration rate.

S: Cuticle.

TWC: Cuticle. Very good. Maybe, one of the plants has a thicker cuticle than the other. Can you feel that? Yes, right. So, there could be a lot of reasons to explain your results. So, that is the way to tackle your discussion. You can suggest what could be the likely habitat of the two kinds of plants. Maybe one of them lives in a dry area while the other needs a lot of water in soil. So, you can try to apply your knowledge, what you called the theory, to explain the result.

5.4 Developing scientific attitudes

Three percent of Hugo's practical teaching was on this aspect. In particular, Hugo wanted his students to report faithfully on what they had done and the results they obtained, as revealed in the following three episodes.

Episode 9-3 – It is your result, you have to interpret it.

S: There is no differentiation between the different coloured spots [which is not what I expect]?

T: I don't know. It is something that you have to consider. It is a fact. It is *your* result. So, *you* have to interpret it.

Episode 9-4 – Explain your expected and unexpected results

S: Sir, it seems that this is not the normal result.

T: You mean this is not your expected result?

S: Yes.

T: But anyway, you got a set of results, right? The fact is that you've got some results, be they accurate or not, okay? You try to record the results first and then try to give some kind of explanation for them.

S: Explain what?

T: Explain your expected results and unexpected results. You said that it is unexpected. Your expected result is not this. So, try to explain why. You should have explanations for that.

Episode 9-5 – Your method should be a genuine report of what you've done

TWC: I have jotted down whether or not you have actually carried out some of the steps like: Did you wash the beetroot discs before putting them into the solutions? … So, if you didn't do this but you wrote down in your report that you had. This is a mistake then… I will check the method section of your report with those in my [assessment] record. Your method should be a genuine report of what you've done. If not, then it is a kind of cheating. Okay?

But at times, a rather different message was delivered to the students. Hugo seemed to regard getting the job done as the first priority of carrying out a practical work. The following is one of the representative excerpts from a lesson:

TWC: The limiting factor for this activity is time [available for completion of the task]… I hope that you can learn from this experience. So, in future, you have to plan to make sure that you really have time to finish the report. Okay? … Unless you believe you have enough time. Otherwise, forget accuracy. So, time to finish the report is the first priority. Okay?

6. ASPECTS OF THE NATURE OF SCIENCE EMPHASIZED

Hugo seldom talked about the nature of science, either explicitly or implicitly. On encountering some unexpected results, Hugo would quickly attribute them to students' improper experiment procedures, as revealed in one of the representative excerpts:

TWC: Hopefully what you've found out would be roughly 3 to 4 times faster in the case of water spinach. Okay? If you've found that the difference is not a significant one, it then reflects some problems in your procedure. For example, you did not dry the surface of the leaf…

7. STUDENTS' ROLE IN THE PROCESS OF LEARNING

In line with his belief in the importance of providing chances for students to "learn a skill of learning how to cope with problems", Hugo often presented the practicals to students as a sort of problem solving exercise. However, he did not expect students to move outside the limits which he had laid down for them; that is, he expected all students to come up with *the* solution he had in mind. That's why he would not provide students with equipment or materials additional to those specified in the lab manual, as revealed in the following representative episode.

Episode 9-6 – You have to plan using only that amount

S: Can I have more glucose solution?

T: Why?

S: Not enough.

T: So you have to plan. Using only that amount… According to the instruction sheet, you are only given 20 ml.

Another striking feature of Hugo's practice was the authoritarian stance he adopted in disputes with students over mark deduction, as revealed in the following episode:

Episode 9-7 – I cannot guarantee that I am one hundred percent accurate but I am fair
It was a post-lab briefing session. Hugo was explaining to the class the TAS mark awarded to each of the students on their abilities to organize and carry out the practical work. One of his students, Sim, was not happy about being accused by Hugo of not "using the white tile as a background against which comparison of colour intensities of the content of different tubes was made":

TWC: Sim. Did not use the stopper… Did not shake content of the tubes before incubation. Did not use the white tile as a background for comparison. [You were awarded] 7 marks [out of 10 marks].

S: No. I did use it.

TWC: You know, when the examiner did not see that you used it (interrupted by the student).

S: No. No. I did use it. It can't be that! It is unfair.

TWC: I am sorry about that. As I have said before, what you have to do is like attending a driving test. You have to grasp every opportunity to carry out those things that you think you have to demonstrate to the examiner. I cannot guarantee that I am one hundred percent accurate. But I am fair.

8. CHARACTERISTIC FEATURES OF ASSESSMENT PRACTICES

Hugo carried out the TAS assessments in a way that was very much the same as administering a public examination. For example, he was as strict as in public examinations in that every student was allowed *exactly* the same time for the task to be completed, as revealed in the following episode (see also Episode 3-1).

Episode 9-8 – Can we start now?
The lab was about to start but some students were having trouble with some equipment and the lab technician was helping them to fix the problems. One of the students, who had no problem with the equipment, asked the teacher:

S: Can we start now?

TWC: Not yet. Not until we are all ready because Mr. Yip [the technician] is still helping some of your classmates to fix their problems.

Hugo was also very strict in not allowing students to look up references during the TAS practicals, as revealed in the following excerpt from a lesson.

TWC: Please *remove all references* from the bench. If I see you refer to this information, I'll regard this as *cheating*.

He was also strict in not allowing students to discuss among themselves, as revealed in the following episode.

Episode 9-9 – I could take this as cheating

Hugo noticed that two students were discussing with each other. He intervened and talked to them in a very serious manner:

T: You two. No discussion. Do you know that I could take this as cheating? You know. When you sit for an external examination, you can't discuss with your neighbour.

T: So, Clarie, next time, when your neighbour asks you question, you'd better refuse to answer her question. This is for your own benefit.

Hugo also avoided giving help to individual students. He would deduct marks from the student if help was deemed necessary, as revealed in the following episode (see also Episode 3-3).

Episode 9-10 – I have to deduct mark from you for the help given

A student was stuck at some point in the middle of a practical. She approached Hugo for advice:

T: Do you think that you need my help?

S: Yes, I need your help.

T: Then, of course, I have to deduct mark from you for the help given. Okay?

S: Er!

T: So, you want to try it by yourself. Or, you need my help?

S: I have no problem with the experimental design. I only have the difficulty with the dilution method.

T: No, I don't think so. It seems that at this stage, your problem is how to determine which one is more concentrated and how much more concentrated than the other. So, you don't have the problem of dilution because you have already finished the dilution process. So, do you think you have to ask for help?

S: I will think about it first. Deduct mark? How many marks are you going to deduct from me?

T: Let me see. For this experiment, err (Hugo is reading the checklist on his notepad) one mark. One out of 10 marks.

The student agreed to the mark deduction. Hugo then asked her to go to a corner of the laboratory where he taught the student the correct method to proceed in a very low voice.

As a result, the lab was very quiet for most of the time. The atmosphere was very much like that in an examination hall.

9. BELIEFS UNDERLYING CLASSROOM PRACTICES

9.1 Beliefs underlying teaching of practical work *per se*

Hugo's emphasis on the procedural side of the practical indicated that he was rather activity-oriented, in that he wanted his students to be able to complete the task on time so as to get satisfactory examination results. The extent of Hugo's concern on this matter can be revealed in the following interview excerpt, during which Hugo was commenting on an instance where in the middle of a practical, a colleague of his came to the laboratory and requested to see a student for some urgent matter. After enquiring about what the matter was about, Hugo begged his colleague's pardon that he could not release the student to attend to the "urgent matter". When asked to comment on the instance during the post-lesson interview, Hugo responded in the following way.

> I have to assess which of the two matters is more important – the student carrying on with her work without interruption or to attend to the "urgent matter". I think the former was more important than what was requested by my colleague. I thought I had to protect the interests of the student – she was attending an assessment which was related to an external examination. Based on this principle, I insisted that the student should continue working instead of attending to the so-called "urgent matter".

Beside this stated belief of "the need to protect the interests of the student", Hugo's practical work instruction was also very much related to his belief about the role of practical work in developing students' self-esteem, as he put it:

> I want to make use of it as a tool for building up students' interest, confidence, or giving them opportunities to deal with their own problem so that they can find out where their own abilities lie. If plan well according to their abilities, this will provide them with a sense of achievement, this can build up their confidence…

As reported earlier, it was this latter belief that had driven Hugo to ask each of his students to carry out an individual ecology project around the campus. In his own words, he thought that this kind of project "is especially important for those who are not good in theory or in English, at least they have a greater chance to have a sense of achievement." Indeed, Hugo highly valued developing in students a sense of achievement through practical work, more so than using it for other purposes. This was also in line with his idea of regarding practicals as a practical experience and his emphasis on "preparing students for future working life". Thus, it is not surprising that Hugo responded as follows, when asked about why he insisted that his students report accurately on their experimental results.

> I told them that I won't give them a low mark for just getting a so-called wrong result… In fact it is an opportunity for them to show off their abilities. I would consider giving them a bonus mark if they can pick up some reasonable discussion from the 'wrong' results.

All in all, Hugo's practical work instruction was very much related to his belief in the role of school education in preparing students for their future working life. And such a belief had overshadowed his other beliefs about the nature of science, at least as reflected in his practical work teaching.

9.2 Beliefs underlying assessment practices

As reported above, Hugo was very concerned about the need to protect his students' interests when implementing the TAS because of the high stakes associated with it. The same reason accounted for the stringent measures he had taken to ensure the fairness of the assessment, as revealed in the following interview excerpt. Here, Hugo was explaining to me why he asked a student to go to the corner of the laboratory and talk to her in a very low voice (see Episode 9-10).

> I didn't want other students overhearing our conversation and thus gaining advantage from it. Otherwise, it would not be fair to her [the student whom I had offered assistance] because I had deducted marks from her already.

In fact, Hugo was caught in a dilemma between facilitating students' learning and maintaining fairness of the assessment, as revealed in the following interview excerpts.

> In fact, I should help students to solve their problems as far as possible… However, due to assessment, I really can't help them.

> Discussion with individual students is very important. However, …it may lead to unfairness because I can only discuss with some students but not all [because of the limited time... In some cases, after doing so, I would tell the whole class in order to be fair.

As implied in the last interview excerpt, sometimes the large class size was perceived by Hugo as one of the main constraints on discussing issues with individual students during the practical. At the beginning of this study,

Hugo had 31 students in his Secondary Six class. There was rarely any individual teacher-student discussion taking place. However, for various reasons, the class size was reduced to 15 when the class was promoted to Secondary Seven. The interaction patterns were very much different from the previous year, with lots of discussions between Hugo and his students. Hugo accounted for the change in interaction pattern with his students in the following way.

> This year differs from last year in that I have far less students in the class. In addition, the duration of the practical is 20 minutes longer. Therefore, I can discuss with each and every one of them individually. In fact, I have planned for that in my lesson planning.

This example shows vividly how Hugo's concern over fairness of the assessment had overridden his desire to help students' learning through more discussion with individual students. Hugo knew that it was good to discuss with individual students but he gave up this idea simply because he had a large class and hence did not have enough time to discuss with each and every student. In short, the concern of fairness took precedence over that of students' learning.

Many of Hugo's assessment actions were also related to his belief in the role of school education as preparing students for their future working life. Thus, he would "deliberately stand in front of the students and check their work", whether he was actually assessing the students or not, so as "to get them accustomed to the situation" as soon as possible. In fact, he expected his students "to grasp the opportunity to demonstrate to the examiner the relevant skills", as he put it:

> This is similar to taking a driving test, if the examining officer does not take note of any of your actions. That's it [and you'd probably fail]. You can't argue with him. The only way is that you have to act wisely and accordingly... That's life.

A similar reason was given by Hugo in explanation of his action of requiring students to observe the deadline for handing in their experimental reports.

> I think this is a kind of training for the students. There are always deadlines attached to jobs that are assigned to you when you are working in the society.

Once again, Hugo's belief in the role of school education in preparing students for their future working life was figuring in his focal awareness when assessing students. In addition, there was a concern about maintaining fairness of the assessment.

10. AN OVERVIEW OF HUGO'S CASE

Hugo saw "preparing students for their future working life" as a prime goal for school education. He held a transmission view of teaching. He saw science as an objective method which can be used to solve problems, although there is no exact procedure to follow. The role of practical work had been transformed to serving his goal of "preparing students for their future working life" rather than goals related to science education.

Both Hugo's assessment activities and his practical work teaching were very much related to his belief that school education should "prepare students for their future working life." There was also evidence of his belief in the unequal power distribution in the classroom, as revealed by his use of the metaphor of "the examiner of a driving test" to describe his role as an assessor. This was also related to his authoritarian style in teaching. His sacrifice of students' learning to maintaining fairness of the assessment was a reflection of his understanding of the TAS primarily as an assessment reform, though he was aware of its pedagogical dimension.

Chapter 10

IVOR

The policeman who fears to be scolded by his superior

1. IT WOULD BE DISASTROUS IF STUDENTS LOSE CONFIDENCE IN YOU

Ivor was a committed teacher who had been working hard to develop his career in the teaching profession. He was trained as a non-graduate teacher. He had served as a part-time biology teacher for about two years in another school before he joined his present school. He taught mainly the lower forms in this school before he obtained his B.Sc. degree after six years of part-time study. He was in his fifth year of teaching A-level biology when this study was undertaken. Though he was already in his late forties, he had just enrolled in a part-time two-year Master degree course in computer programming. Ivor hoped that, upon graduation, he would be given more opportunities to teach computer studies, if not to quit teaching biology completely. He viewed biology as distinctly different from other science subjects, in the following way.

> In learning other subjects like chemistry and computer studies, you must understand the preceding modules before you can understand the next one. Biology is not the same... Your not understanding the previous lesson would not affect your understanding of the next lesson...

B.H.W. Yung, (ed.), Assessment Reform in Science, 129-140.
© 2006 *Springer. Printed in the Netherlands.*

Ivor's view of biology knowledge as being disjointed parts that are unrelated to each other might explain why he regarded his six years' study in the part-time biology degree course as not contributory to his teaching at all.

Ivor believed that the most important thing for a teacher was a good teacher-student relationship. According to him, if the relationship is good, any problem can be solved. However, he noted that such a good relationship had to be built upon students' confidence in the teacher. In his words:

> If students find that the teacher is not knowledgeable, it is impossible for the teacher to build up a good teacher-student relationship. Students can tell quickly if a teacher is knowledgeable or not after listening to his teaching for a few minutes… It would be disastrous if students lose confidence in you.

To Ivor, a knowledgeable teacher meant, besides other things, "to know the syllabus and the trend of examination questions very well". Related to such a view was Ivor's belief in teachers as the authoritarian source of knowledge and his view of teaching as a transmission of knowledge. As a corollary, Ivor also did not believe in peer learning pedagogy.

> If the student wants to ask a question, he should ask the teacher and not his classmates. His classmates may not give him a correct answer… Learning is mostly guided by the teacher. Learning alone won't achieve much. The progress of the lesson is not controlled by the student. It is mainly controlled by the teacher… If I can provide them with good and concise notes, students can learn better…

2. A COMPLETE SCIENTIFIC METHOD SHOULD COMPRISE FORMULATION OF A HYPOTHESIS

Ivor's beliefs about science were vague and fragmented. When asked about what science is, he drew on some textbook definitions that "science is a systematic study which requires people to think in a scientific way; to make observation, to postulate hypothesis, to test, to make comparison… Just carrying out some operations is not science… Science is to have a questioning attitude and to ask as many questions as possible." In Ivor's view, "a complete scientific method should comprise formulation of a hypothesis, testing and discussion, etc."

3. THE ROLE OF PRACTICAL WORK

Ivor saw the role of practical work as more important in terms of training students in acquiring scientific method than in verifying the theories taught. He also emphasized using practical work to arouse students' interest. This fitted in with his belief that students would learn things fast if they were interested. He added that this would, however, depend on the nature of the practical itself, and that not every practical could be made interesting for students. Thus the reality for his students, as described by Ivor, was often a feeling of "being torn into pieces on completion of the practical". This was because although Ivor wanted his students to learn with interest, he also believed that "students will not be serious about it if you tell them that a practical will not be assessed. They will not learn if they are not serious." So, in fact, it was this latter belief of "examination-driven learning" rather than belief in "interest-driven learning" that was guiding Ivor's classroom actions–for example, his decision to inform the class at the beginning of the term that all practicals would be assessed throughout the year. In his words, he wanted "to push them to perform better… to get a higher mark for the report".

4. CHARACTERISTIC FEATURES OF PRACTICAL WORK TEACHING

Amongst the ten teachers in this study, Ivor was the second teacher who was most concerned with the assessment-related issues. Of the T-S interaction on practical work teaching, 55% was focused on the procedural aspects, 24% on concepts underlying scientific research, 4% on biology theories underlying the practical and about 2% on scientific attitudes. In other words, Ivor emphasized training students to acquire basic laboratory skills and there was little discussion of biology theories related to the practical work or aspects of scientific attitudes.

Regarding the pattern of T-S interaction, 23% were directions from Ivor, 18% were on inquiring about students' progress of work, another 23% involved Ivor answering students' questions by telling them the answer directly, while 34% involved Ivor guiding students to the answer.

4.1 Developing basic scientific skills

Ivor emphasized training students to acquire basic laboratory skills. He wanted them to work in a systematic and safe manner inside the laboratory, as revealed in the following episode:

Episode 10-1 – Laboratory procedures and safety measures

It was a practical on estimation of reducing sugar content in a piece of grape. The following were instructions given by Ivor to his students at different times of the practical on the laboratory procedures and safety measures, etc.:

During the pre-lab briefing:

TWC: Number 5, set up a boiling water bath but to do it in a safe manner. You have to control the size of the flame… Also make sure that the mouth of the test tubes are not pointing at somebody...

At various points while he was circulating around the lab:

T: Hold on. Hold on. When you put the dropper on the bench, and then into the glucose solution, this may contaminate the glucose solution. Did you notice that there is a test tube sticking at the side of the bottle for holding the dropper.

T: Adjust the flame to a smaller one first, until you are going to heat the solution.

T: Why is your test tube pointing outwards (towards someone)?

S: Sir, how much stock solution should I get?

T: If I answer your question, I will deduct marks from you. You have to decide for yourself. In fact, I should ask you instead. How much do you think you should obtain from here?

S: 0.001 ml.

T: How could you obtain such a small amount of solution from the stock? That is nonsense. I really want to ask you, "How much do you want to obtain?" Hold on. Don't get any solution yet. Go back to your own seat and plan it first. Don't just pour out the solution without knowing exactly how much you want. Think about it first and then explain to me before you come out to get the solution again.

After a while, the student returned to the teacher bench to get the stock solution.

S: I want to get 10 ml.

T: Why 10 ml?

S: Because I have to make the dilution.

T: Come over here. Look at your dilution table first… So, here, you can do some calculations. From the table, you can see, right, here. This is the volume of stock solution. These all add up together to make about 6 ml only.

S: So, I just get 6 ml.

T: Should we not work out the rough amount that we require, and not just make a wild guess?

At times, Ivor could be quite "picky and fault-finding" of his students' performance, in accordance with his description of himself when assessing his students. The following is one of the representative episodes in this area.

Episode 10-2 – How thick are these beetroot discs?

T: How thick are these beetroot discs?

S: 1 mm.

T: 1 mm? Did you measure it with a ruler?

S: They are all of the same thickness.

T: That means how thick are they? How many mm?

S: 1 mm.

T: Without any ruler, how do you know that it is 1 mm?

S: My ruler is rusted. (The student replied in a rather impatient manner.)
T: What did you rely on to tell me the thickness of the discs?
S: They are all of the same thickness!!
T: Yes, they are of the same thickness but they are not necessarily 1 mm thick.
S: I measured one of the discs and then assume that the rest of them are all of the same thickness.
T: Where is the ruler? Show it to me.
S: The ruler, of course, is inside the pencil box.
T: How come that you know that it is 1 mm even without using a ruler?

4.2 Enhancing understanding of the concepts underlying scientific research

About 24% of Ivor's practical instruction was related to teaching concepts underlying scientific research. Below is one of the representative episodes in this area.

Episode 10-3 – The importance of setting up controls
Ivor was discussing with a student some unexpected results from an experiment investigating protease activities in different kinds of fruit juices:
S: I have added buffer solution to this one but not this one.
T: There is a clear zone for the treatment with buffer solution only and no clear zone for this one. That means, there is some problem with the buffer. This is due to the effect of the buffer [instead of the fruit juice as expected]. So, if you had not set up the control with buffer only, you would have drawn a wrong conclusion. It is lucky that you have set up the controls.
In the post-lab debriefing, Ivor reiterated the importance of setting up controls in scientific investigations:
TWC: From this practical, you should have appreciated how important a control is. Imagine, if none of you had set up a control using buffer alone, we may never know that the buffer solution alone can have a proteolytic effect instead of the fruit juice itself. Right? The conclusions can be very unreliable. That's why, if we don't have a valid control, any conclusion can be invalid.

4.3 Aspects related to scientific attitudes

About 2% of Ivor's practical teaching was on this area. He mainly asked his students to be truthful about their experimental results. But he seldom discussed with them why they had to do so, as revealed in the following excerpts:

S: Sir, why is the result like this?
T: Just *report faithfully* what you did. I can't teach you this.

However, at times, Ivor was hinting to students that the prime purpose of carrying out practical work was to get results in order to generate a decent report for assessment purposes, as revealed in the following excerpts from two of the lessons:

T: It would be *troublesome* if you don't get any results.
T: *Put more effort in writing up your report.* Do it better and in more detail. Write up to five pages. Five full pages.

5. ASPECTS OF THE NATURE OF SCIENCE EMPHASIZED

There was no explicit discussion of any kind about the nature of science in Ivor's teaching of practical work though, at times, he could have made use of certain instances that arose during the course of teaching. Nevertheless, given the great emphasis that he had put onto developing students' basic laboratory skills, he should have conveyed the message to the students that scientific investigations have to be systematic; it is about accuracy and logical thinking, etc.

6. STUDENTS' ROLE IN THE PROCESS OF LEARNING

As evident in episodes presented above, Ivor was often ordering or giving directions to students about what they should do, despite his efforts to set the practicals in the form of some questions to be solved by the students. Thus, students were expected to be followers of the teacher's instructions in order to get the job done in a fast and efficient manner, as revealed in the following excerpt:

Episode 10-4 – stirring a solution with a glass rod

T: You don't have to swirl the tube to mix the content inside now. Swirl it only when you are going to use it. And swirl it like this, using a glass rod.
S: But then some of the solution will stick onto the glass rod [and hence it will affect the results of the experiment].
T: You just leave the glass rod there [in the solution] through out the experiment.
T: In fact, it doesn't matter whether there will be some sticking onto the glass rod or not. This is because you will be using just about half of the volume of the prepared extract. The remaining half is useless and will be discarded.

7. CHARACTERISTIC FEATURES OF ASSESSMENT PRACTICES

Ivor's assessment practices were very much the same as those which one can find in a public examination hall. For example, he was the most concerned, amongst the ten teachers in this study, about the need to supervise students' assessed work, as revealed in the following instance:

Episode 10-5 – Supervision of students' work
One day, Ivor had to be in charge of a detention class after school. However, his Secondary Six students could not finish writing up their reports on time and hence the practical had to extend beyond normal school hours. Upon Ivor's repeated urge to work faster, his students suggested to him to delegate the responsibility of supervising students' work to the lab technician so that he could be released for the duty of looking after the detention class. But Ivor refused and insisted on staying behind to supervise their work until they finished. Instead, he asked one of the students who had already handed in the report to act for him in taking charge of the detention class.

When asked to comment on this episode, Ivor equated the situation to that of invigilating a public examination, saying that "the examination has to be cancelled if there is no teacher invigilating the examination". This also explained why Ivor did not allow students to finish writing up their reports at home, as some other teachers occasionally did.

Ivor was also very serious about the issue of giving exactly the same time allowance for each and every student to complete the assessment task, as revealed in the following episode.

Episode 10-6 – Drawing a red line on the lab report
The lab was about to start. Students were organizing their workstations. Ivor was circulating around the lab to ensure that no students started working before they were told to so:
T: You can organize the apparatus but do not start the experiment until I tell you to do so.
T: Do not start labelling the tubes. It has not started yet.
T: Don't start the experiment until I tell you to do so.
Towards the end of the practical, Ivor made the following announcement to the class:
TWC: Today, the deadline for submission of report is 3:30. No delay, no delay. Hand in your report on or before 3:30.
At 3:25 p.m., Ivor made another announcement:
TWC: You have 5 minutes left.
At 3:29 p.m., Ivor reminded the class again:
TWC: You have one minute to finish your conclusion and then submit your report.
At 3:30, when the time was up, Ivor said to the class:
TWC: Pens down.
He then circulated around the lab to stop students from writing any more. To one student he said:
T: Stop and write no more otherwise I will not accept your report.

S: Can I just finish this sentence?
To another student who was still writing, he said:
T: I am not kidding. I won't mark those [that you wrote after the time's up signal].
S: I just want to finish this sentence.
T: No, you can't.
Ivor continued to circulate and drew a red line on the students' reports so as to indicate where the reports were up to when the time was up. At the same time he said to the class:
TWC: Come on. Hand in your report now. Put it on the table fast. I won't mark the part which is beyond the red line.

During the practicals, Ivor also refrained from offering help to individual students, as revealed in the following episode.

Episode 10-7 – I can't tell you
Ivor was circulating around the lab. One student approached him and requested help on how to make use of the collected experimental results to compute the concentration of glucose in the unknown sample, as required by the lab sheet:
S: I added 4ml .... How many times of dilution have I made here?
T: I can't tell you now.
S: But, I don't know how to do the calculation!
T: I can't tell you. If I teach you now, then I can't give you the marks for this part.
S: (?)
T: Just report faithfully what you did. I can't teach you this.
S: (?)
T: I have told you already. I can't tell you any more.
S: Come on, please.
T: No, I can't.

One characteristic feature of Ivor's assessment action was that he often pointed out to students the mistakes that they had committed on the spot, in addition to telling them that marks had been deducted from them too, as shown in the following episode.

Episode 10-8 – I can't give you mark on this
T: Your method works only if the final volumes are the same. But the problem is, can you add water into them such that the final volumes are the same in all the 4 tubes. Can you? No, you can't.
S: What do you want me to do then?
T: You have already committed a mistake here. If I did not alert you of this, I would not know that you are reporting the results using the wrong method, I am telling you that I can't give you mark on this because you are using the wrong method here. These results are meaningless.

Ivor also informed his students at the beginning of the term that he would be assessing them in all practicals, which he did. At times, he also made explicitly clear to his students the assessment criteria, as revealed in the following excerpt from a lesson.

TWC: I will also be assessing you on the following practical skills. Number 1, good handling of large numbers of test tubes… You have to label all containers before you use them. You have to arrange them in an orderly way… Number 6, making good serial dilution… Work according to a dilution table… Make sure that you swirl the mixture thoroughly…

So, in a way, Ivor was both teaching and reminding students of the proper techniques while highlighting to them the relevant assessment criteria. After the students had completed the practical and handed in their reports, Ivor would hold a very short post-lab briefing, highlighting the areas on which marks had been deducted from the students, as revealed in the following excerpts from two of the lessons:

TWC: Marks had been deducted for looking up reference materials when writing up the report. Marks had been deducted for asking for help from me on how to write up the report.

TWC: Marks had been deducted for helping your neigbours to do their work. Next time, don't ever try to prepare the dilutions for your neigbours. Tell me before you do it. You need my permission before you help others.

Overall, assessment activities constituted a prominent feature of Ivor's practical work instruction.

8. BELIEFS UNDERLYING CLASSROOM PRACTICES

8.1 Beliefs underlying teaching of practical work *per se*

Ivor's emphasis on training students to acquire basic laboratory skills and to understand the concepts underlying scientific research matched very well with his view of role of practical work in school science education; that is, practical work is more important in training students on scientific method than on verifying the theories taught. Particularly prominent was his belief in the importance of helping students to get better examination results. This explains why he stopped asking his students to carry out individual project work once the TAS did away with such a requirement. In his words, "I am sure that project work would take up a lot of time for students... It is not to their benefit given the marks awarded to the project… I will not carry it out now." The very same reason explains why he assessed his students in nearly every practical. As he put it:

> For the sake of the students, it is better to assess them on more occasions.

This was because Ivor believed that exerting a bit of pressure on students could make them work harder in their studies. He said, "If I tell them that I am not going to assess a particular practical, the students will not be serious about it… They won't learn if they are not serious." So, in fact, it was this latter belief in "examination-driven learning" that was guiding most of Ivor's classroom actions.

8.2 Beliefs underlying assessment practices

Many of Ivor's assessment actions were related to his perception that the TAS was a part of the high-stakes HKAL Biology Examination and that teachers were held accountable by HKEA to assess the practical competence of their own students. Words like "fear", "worry", "HKEA would not believe", "not allowed" and "cheating" predominated in his interview excerpts. The following interview excerpts help to illustrate Ivor's various concerns in trying to cope with the TAS requirements and its many restrictions:

> I won't do this... The TAS does not allow the teacher to conduct a post-lab discussion before students are asked to write up their own report. Otherwise, it is just cheating.

> I fear that the coordinator would say that my marking is too lenient… I'd rather deduct marks from my students without any special reason… I add one additional assessment criterion to my checklist – that is, overall performance. With this criterion, I can deduct marks from students relatively more easily... You know. I have to keep their marks low.

Thus it is not surprising to find that Ivor had taken the trouble to compile a lot of assessment records "just for scrutiny by the TAS coordinator". This focus on putting it "in writing" was derived in large part from his perception that the purpose of the exercise was increased accountability. The perception of the need to "keep the students' marks low [in order that HKEA would not query]" had also forced Ivor to adopt a "picky and fault-finding" attitude when assessing his students, even though he knew that it was unfair to his students.

Related to his perception that the TAS is a part of a high-stakes examination, Ivor was also very concerned that he had to be seen as a fair assessor by his students. For example, his intention to minimize disputes over mark deduction by pointing out to students their mistakes on the spot was clear in the following interview excerpt.

> If I can point out the mistake to them on the spot, they cannot challenge me later on about why certain marks have been deducted from them... This is especially the case when a substantial amount of marks have to be deducted... Otherwise, this may worsen the student-teacher relationship. No student will be happy to have marks deducted.

The very same reason explained why Ivor would, at times, invite his students to consult their classmates who were performing very well, as he put it:

> If not, other students will not know how poor their own work is. Otherwise, they will query why they get such a low mark. If they have a chance to look at the good work of others, then they will not complain any more [about the low marks that I awarded to them].

As evident from above, Ivor's main concern was to gain students' confidence in his judgment on assessment issues. This also explains why Ivor required his students to start working and to finish off the practical at *exactly* the same time. The same reason accounts for his act of drawing a red line on the reports of those students who did not stop writing after the time-is-up signal. In Ivor's words, "Otherwise, it would "not be *fair* to the obedient students".

Ivor was also concerned with the inability to differentiate between students if he were to carry out a discussion with the whole class on their experimental results before they were asked to write up the report individually. This was because he held a perception that students would be *told* '*everything*' during the discussion. Similarly, he was caught in a dilemma between facilitating students' learning and keeping to the TAS regulations.

> Of course, on-the-spot discussion is better if we are to look at it from the teaching point of view. I just worry if it is cheating or not...

To sum up, Ivor's assessment practices were very much related to (1) his concerns about the issue of fairness, (2) his perception of the role of the TAS in differentiating between students as accurately as possible and (3) his belief that he was held accountable by the public examining body for assessing his own students. Of course, he was also aware of his role, as a teacher, in facilitating students' learning. However, in most cases, assessment considerations superseded those of pedagogical ones.

9. AN OVERVIEW OF IVOR'S CASE

Ivor had a weak sense of what science is and the role of practical work in science education. He viewed teaching as a transmission of knowledge, with the teacher as the authoritarian source of knowledge. He believed in the effectiveness of "examination-driven learning" more strongly than "interest-driven learning".

Ivor's emphasis on training students to acquire basic laboratory skills and to understand the concepts underlying scientific research matched very well with the way he saw the role of practical work in school science education. That is, practical work is more important in training students on the scientific method than on verifying the theories taught. However, figuring in his focal awareness was his belief in the importance of helping students to get better examination results. This was aggravated by his perception that the major role of the TAS was to differentiate between students as accurately as possible, and his concern about maintaining fairness of the assessment. Of course, he was also aware of his role, as a teacher, in facilitating students' learning. Disappointedly, however, in most cases, assessment considerations had taken over those of pedagogical ones. In other words, the TAS was understood primarily as an assessment reform.

Chapter 11

JOHN

The examination-driven teacher

1. MY WORK IS ALWAYS DRIVEN BY EXAMINATIONS

John was older and more experienced than most of the other teachers in this study. He was in his late forties. He had taught A-level biology for 16 years before he emigrated to Canada, where he taught in a public school for two years. On his return to Hong Kong, he joined a prestigious school. It was during the fourth year of his teaching in this school that the present study was carried out. Indeed, John took pride in the contribution of his teaching experience to his teaching.

> If I hadn't had any experience, I would not have been able to produce such a year plan. With the experiences I have gained over the years, I can modify the plan bit by bit, and year after year… The most important point is that I can enrich my own repertoire so that I can make adjustments easily, even if the nature of the student intake changes…

John put a lot of emphasis on preparing his students for public examinations – a goal which he perceived to be supported by his students. Anything seen to be outside the syllabus boundaries by him would be given peripheral status. The extent to which the public examinations or the TAS had affected John's teaching in general and the teaching of practical work in particular may be seen in the following excerpts.

B.H.W. Yung, (ed.), Assessment Reform in Science, 141-150.
© 2006 *Springer. Printed in the Netherlands.*

> My work is always driven by examinations. Being a biology teacher, you need to have a thorough understanding of the subject matter, know the syllabus well, highlight for students the key points, ... teach them how to study, how to answer the examination questions.
>
> There are several things that one would look for when one is assessing a teacher's performance. First, how do the students evaluate your teaching? Second, your students' external examination results, maybe. Third, other fringing activities like whether he is a marker or a member of the subject committees in the public examining body.

2. THE ROLE OF PRACTICAL WORK

The impact of his concern about good examination results was not only manifested in teaching related to the theory examinations but also in John's teaching of practical work, as revealed in his description of how he prepared his students for practical work at the beginning of the term.

> I devoted more than four lessons to the "concepts of evidence" using exercises from overseas past examination papers... This is necessary because the TAS is asking for this. There is also a possibility that this kind of concepts will be examined in the theory examination.

As illustrated here, John did not look at the necessity of teaching students the "concepts of evidence" from a pedagogical point of view, or from the view of developing students' scientific literacy. Rather, he saw this in terms of satisfying the examination requirements. Such a preoccupation had also affected the way he saw the role of practical work in science teaching, as revealed in how he selected practical work for his teaching.

> The practical work must fit the syllabus... This can supplement the theory they have learnt... In addition, those report writing and graphical skills can help them in the theory examination... The practical work should be able to produce clear-cut, accurate and reliable results so that it will be easier for the teacher to assess the

> accuracy of the results obtained by the students. That is, the ability of the practical to differentiate stronger and weaker students reliably and accurately is certainly one of the considerations for its inclusion in the year plan.

Thus, for John, practical work was mainly a tool for assessing and differentiating his students' practical competence. It was also seen as a supplement to theory teaching and preparation for the theory examination.

3. SCIENCE IS DIFFICULT TO DEFINE

For John, science is "difficult to define… Certain topics are science because they are included in the science syllabus… Science has its own scope… It is defined over the years. There is a consensus... Secondly, it depends on the kind of activities… Classification is a kind of activity in science… Calculation is science… Observation is a skill which is necessary in science training… Asking questions [about a phenomenon] is science."

4. TEACHING IS CONVEYING INFORMATION

John held a transmission view of teaching, as revealed in the following interview excerpt:

> Conveying information is teaching… Describing the steps of doing a calculation is teaching… Some teaching… can be unidirectional like TV programme, or following a recipe to make bread.

5. CHARACTERISTIC FEATURES OF PRACTICAL WORK TEACHING

John was the teacher in this study who least interacted with the students during the TAS practicals. He was the teacher who was most concerned with assessment-related issues. For John, the prime goal of practical work teaching was to serve the purpose of assessing students' practical competence. As such, he was rather activity-oriented. That is, most of his interactions with students were on the procedural side of the practical (45%) and relatively little time was spent relating the practical to underlying biology theories (8%) or developing students' scientific attitude (3%). Regarding the pattern of T-S interaction, 31% concerned inquiring about students' progress of work, 18% comprised directions from John, another 24% constituted answering students' questions by telling them the answer directly while 19% involved John guiding students to the answer.

5.1 Developing basic scientific skills

Of the relatively small amount of T-S interaction during the TAS practicals (132 DTUs per lab session), a large proportion was spent on the procedural side (45%), as revealed in the following excerpts from various practicals.

T: Next time, don't put these things here. Throw them away immediately after use. If not, marks will be deducted. Okay?

T: This is called overheating. Don't do this again. One more thing, you should put the mat below this, below the burner.

T: Just do it quickly. No need to be so accurate.

John's focus on procedural matters was evident in all these excerpts. His underlying concern was to get things done properly, safely and quickly. His assessment-oriented attitude is evident from the first excerpt; his concern with helping students to excel in their examination is implied in the last excerpt.

5.2 Understanding of the concepts underlying scientific research

About 15% of John's practical teaching was related to teaching concepts underlying scientific research. Below is one of the representative episodes.

Episode 11-1 – How many controlled variables are there in this experiment?
John was having a pre-lab briefing with the whole class on a practical investigating the effect of different chemicals on the membrane permeability of beetroot tissue:
TWC: What is the dependent variable in this case?
S1: The colour intensity.
TWC: Yes, it is the colour intensity. What is the independent variable then?
S2: The different chemicals.
TWC: Chemicals. Yes. The tricky thing in this experiment is the controlled variables that we should be very careful of. There are a number of controlled variables. If you look at the manual, … the time period is a controlled variable. And so is volume, ... etc.
TWC: So, you better think about this very carefully. How many controlled variables are there in this experiment?… You'll find that if you're not controlling those parameters for all the chemicals, obviously it will affect your results.

5.3 Developing scientific attitudes

About 3% of practical instruction fell into this category. All were concerned with asking students to report their experimental results faithfully in their reports, as revealed in the following excerpt.

S: It doesn't change its colour [as expected].
T: Then, just record what you observe.

As evident from above, there was no explicit discussion in relation to the nature of science for why the students should report results faithfully. In addition, John was sometimes more concerned with students' assessment marks than developing in them the appropriate scientific attitudes, and hence he adopted a more relaxed stance on whether students should report on the actual results or not, as revealed in the following excerpts from one of the lessons.

S: Should I report the actual results or the expected results?
T: You have to decide for yourself.

6. ASPECTS OF THE NATURE OF SCIENCE EMPHASIZED

There was no explicit discussion of any kind about the nature of science in John's teaching of practical work though, at times, he could have made use of certain instances that arose. Below is one such example. Because of the unexpected results, John had to introduce students to other possible methods of collecting the "out of range" data in addition to the quantitative method that he had suggested at the beginning of the practical.

TWC: It doesn't matter if you find that the quantitative method of assessing the colour intensity of the solutions did not work with your set of results. It doesn't matter. You can use this method, the semi-quantitative method. Okay? Or, if you think that it is no good, you can use this [the qualitative method]. It's up to you. Okay?

However, John stopped short of making use of this episode to discuss with the students the dynamic nature of the scientific investigative process.

7. STUDENTS' ROLE IN THE PROCESS OF LEARNING

John did not expect his students to deviate from what he had planned for them in carrying out the practical work, as revealed in the following excerpt:

S: Can I have more test tubes?
T: No. Just 7. There is a reason for this. Think more carefully about it.

He didn't care to probe into the reason why this student wanted to have the extra equipment to carry out her practical work, as some other teachers in this study have done. John's practice of giving students a summary remark sheet on returning the marked reports to them was reflective of his view of students as passive recipients of knowledge, as revealed in the following interview excerpt.

> Students can analyze it better at home if I give them the summary sheet. The remark sheet in fact is a very comprehensive summary which I have been compiling when I am marking their reports. I feel that basically they should have no problem after reading the summary sheet. That is, they should have no further queries or questions that they want to ask.

Thus, for John, the summary sheet represents an authoritative and comprehensive source of scientific information which could be easily assimilated by the students after reading it on their own.

8. CHARACTERISTIC FEATURES OF ASSESSMENT PRACTICES

John's assessment practices were very much the same as those which one can find in a public examination hall. For example, he tended to standardize every minute detail of the assessment setting so as to maintain fairness of the assessment, as revealed in the following episode where he was trying to delay the start of the practical until all students had arrived:

Episode 11-2 – Don't start until everyone is here
The lab was about to start. Students were organizing their workstation and John was making a head count to see if all students were present. He noticed that one of the students, Catherine, had not arrived yet:
TWC: Organize yourself but please don't write up your report. Don't start writing up the report yet.

After waiting for another two minutes, Catherine still did not show up:
TWC: Now, everyone is here except Catherine. What has happened to her? Is she coming?
S: No, she is not coming. She is absent today.
TWC: Okay then. Everyone is here. So, we can start now.

John was conscious of not allowing students to discuss among themselves, nor did he allow students to help each other in carrying out the practical work, as revealed in the following episodes.

Episode 11-3 – Please work on your own
John noticed that one student, Annie, was helping her neighbour, Mary, to fix some problems with the experimental setup. He intervened:
T: Sorry, please work on your own. Don't work together. What's wrong with you, Annie?
S: Mary asked me to help her.
T: No, no. You can't. She has to do it on her own.

John also avoided discussion with individual students and declined to offer assistance to them, as revealed in the following two episodes:

Episode 11-4 – I fear that I will do it wrongly
A student was not very sure about her own experimental design. She came to John hoping that she would get some advice from him:
S: Should I take out the beetroot discs from the chloroform and paraffin oil and put them into these two tubes. And then ...
T: Think about it yourself.
S: I fear that I will do it wrongly.
Having got no answer from John, the student looked very puzzled and looked around to see what others were doing.

Episode 11-5 – I don't want to answer this question
S: I cut this into smaller discs to increase the surface area and then... What do you think?
T: I don't want to answer this question.

Indeed, the classroom atmosphere was very much like that in examination halls (see also Episodes 13-1 to 13-6).

9. BELIEFS UNDERLYING CLASSROOM PRACTICES

9.1 Beliefs underlying teaching of practical work *per se*

Much of John's practice in his teaching of practical work was related to his concern about preparing students to obtain good examination results, as reflected in his saying, "My work is always driven by examinations." John's primary goal, for which he spent more than four lessons teaching students about the "concepts of evidence", was to prepare students for the theory examination. He did not look at this teaching from a pedagogical point of view or from the view of developing students' scientific literacy. His selection of practical work for inclusion in his teaching was also examination-oriented. That is, whether the practical work can supplement the theory taught, whether those report writing and graphical skills could help students in the theory examination, and whether the practical work will be able to differentiate students reliably and accurately.

Thus, for John, practical work was mainly a tool for assessing and differentiating his students' practical competence. It was also seen as a supplement to theory teaching and preparation for the *theory* examination. Sometimes, John's concern with helping students to obtain good examination results would override his beliefs about the nature of science and the role of practical work. For example, John admitted that he was caught in a dilemma between inculcating a good scientific attitude in students and helping them to get good examination marks. He hesitated when asked to comment on this issue:

> Err... It is good for them to report but it is no good for their TAS marks because they are inaccurate.

Lastly, John's practice of giving students a summary remark sheet on returning the marked reports to them was reflective of his view of teaching as a transmission of knowledge from the teacher to the students.

9.2 Beliefs underlying assessment practices

Underlying John's assessment activities was his belief that the TAS required him to differentiate his students' performance in practical work as "accurately" as possible. Thus, it was essential for him to avoid interacting with students or offering assistance to them during the practical in order that they would not "get very similar marks". The very same belief underpinned John's decision not to carry out lengthy pre-lab and post-lab discussions with his students, as he put it:

> You simply can't have a pre-lab discussion with the students. You can't tell them too much… The major reason is that I can't assess them if I discuss the practical with them. There is no way that I can differentiate their performance.

Underlying John's practices of avoiding discussion with individual students and not offering assistance to them was his concern about maintaining fairness of the assessment.

> You have to be fair and accountable. This is what I should have done. Not because of the students, though they are a bit more concerned about their marks.

In particular, he wanted to be seen as a fair assessor by his students, as revealed in the following interview excerpt.

> Discussion with individual students is good but … the other students will complain that you are teaching a certain student. They will think in this way. So, I am very cautious of this because they take this very seriously… If I answer a student's questions, I am sure that I will have to tell the rest of the class using the microphone.… It is not fair unless you are going to discuss with every other student and that you discuss with them more or less the same questions.

As evident from above, John perceived fairness as equivalent to each student receiving exactly the same treatment from the teacher during the assessment. This included the kind of help each student received when discussing with the teacher individually. John thought that this kind of help was very difficult to quantify and hence the best way to maintain fairness was not to discuss with individual students at all.

The second factor that inhibited John from discussion with individual students was his fear of being perceived by students as 'unfair' in teaching some students but not others. The need to be seen as a fair assessor is also evident in the following interview excerpts.

> Why doesn't the teacher answer the student's question directly [but referring her to another student with whom the teacher had discussed a similar problem]? It would only give the student a feeling that you don't want to answer her question… I fear that the student will go home and complain to her mother that I didn't answer her question. I would then be in trouble.

> Referring students to consult a classmate who is performing well is important because it tells students that the standard of work could be as good as what they see; and that I do not fool them…
>
> Just marking their reports seems not very accurate. Giving them model answers would give them an impression that I am very fair in the process [i.e. marking with reference to some objective criteria].

As evident in these excerpts, there was a lack of trust between John and his students to such an extent that he had to rely on "black and white documents" (of criteria for assessment) to make students believe that he was really fair in carrying out the assessments. Indeed, he was amongst those teachers who were most conscious of the need to be seen as a fair assessor by the students they taught. His main concern was to gain students' confidence in his judgment on assessment issues.

10. AN OVERVIEW OF JOHN'S CASE

John was very examination oriented. He held a transmission view of teaching. He had an amorphous view about the nature of science and he saw practical work mainly as a tool to differentiate students' practical competence.

Much of John's teaching of practical work was related to his belief in the importance of preparing his students for obtaining good examination results. This, in turn, was related to how he saw the role of practical work in science teaching. That is, it was mainly seen as a tool for assessing and differentiating between students' different levels of practical competence "as accurately as possible". Furthermore, it was seen as a supplement to theory teaching and preparation for the theory examination. Its role in developing students' scientific attitudes, and the like, were given lower priorities. In other words, the TAS was understood primarily as an assessment reform. This explains his concern about the need to maintain fairness of the assessment and the need *to be seen* as a fair assessor by his students. His belief in the transmission model of teaching was also evident in his teaching.

Chapter 12

LOOKING ACROSS THE CASES

A preliminary analysis

Chapters 4 to 11 presented the case stories of eight of the ten teachers studied in the project. This chapter draws upon the data described in the previous chapters, but also that of the other two teachers studied, Fay and Glen, to give an account of the teachers' TAS practices and their associated beliefs with reference to four specific foci. After examining each of the foci, this chapter uses this initial analysis to suggest further in-depth analyses of three aspects of the data.

By looking across the cases, it is possible to identify a number of common themes that provide useful markers by which to compare and contrast the teachers' TAS experiences and their associated beliefs. Mostly these relate to the questions that either initiated the study (e.g. What are secondary science teachers' beliefs about science, teaching and learning?) or emerged through the ongoing analysis of the data. An example of the latter is the teachers' concern about fairness, both in their classroom actions and in our conversations. These concerns have prompted me to conduct a more in-depth analysis of the data to determine exactly what the teachers meant when they talked about fairness (see Chapter 13).

1. TEACHERS' BELIEFS AND ASPIRATIONS

One of the aims of the study was to find out the teachers' beliefs about science, as well as their beliefs about teaching and learning. Table 12-1 provides a summary of the findings in tabulated form to facilitate cross-case comparison. This brief but concise summary also serves as a quick and convenient reference for readers to refresh their memories about the teachers, the beliefs they hold and what they aspire to achieve in their classrooms.

B.H.W. Yung, (ed.), Assessment Reform in Science, 151-163.
© 2006 *Springer. Printed in the Netherlands.*

Table 12-1 Teachers' visions / aspirations and major educational beliefs

Teacher	Personal experience/ vision / aspirations	Beliefs about science and science teaching	Beliefs about role of practical work	Beliefs about teaching and learning
Alan	• Not just teaching for making a living • Teaching contributes to life – student's life • Educating students to become good citizens • Facing problems is the correct attitude for living life	**Dynamic / theory-laden view:** • Science is to seek understanding about nature – the stable, ultraistic, existential facts • The greatest thing about science is its predictive capability • Science is not all mighty; science has its own limitations • Scientific quest is a *never ending* business • Successful science learning is learning how to make knowledge, how to think and how to learn	**Affective / cognitive / motor skills:** • Cultivate a "pragmatic attitude" towards science • Cultivate the attitude of objectivity in doing science • Cultivate "a thinking habit" which will help Ss solve problems for themselves • Practise problem solving, develop conceptual understanding and rectify misconceptions	**Learning-focused:** • A teaching philosophy of "bring it low, spiral it up" so as to help remove from Ss the feeling of learning "alien" ideas • Learning has to begin with some ground work upon which Ss construct their own knowledge • Learning as a process of "gradual self-understanding" not guided by the teacher • Teaching has to be "dynamic" and involve interaction with students • Peer learning pedagogy is good for students' learning
Bob	• A teacher with a mission who evaluates his own teaching and hopes to improve continuously • Cares about the Ss in all aspects of their development • Stands firm on things which he thinks are beneficial to Ss	**Dynamic / theory-laden view:** • Scientific method is central in the development of science but there is no one single scientific method • Science is *dynamic* and *on-going* • Observation is *theory laden* • Science seeks truth but it is not able to arrive at "The Truth" • The highest level of science teaching is teaching the thinking underlying scientific method	**Affective / cognitive / motor skills:** • Inculcate in Ss a scientific attitude • Enhance Ss' understanding of the scientific method and its underlying thinking • Enhance Ss' problem solving ability	**Learning-focused:** • Good teaching is good learning • Teacher is just a facilitator; much of the learning has to be carried out by the Ss themselves; the more they think, the more they learn • Individual project work contributes significantly to students' learning

Table 12-1. Continued

Teacher	Personal experience/ vision / aspirations	Beliefs about science and science	Beliefs about role of practical work	Beliefs about teaching and learning
Carl	• Drilling Ss for examination is a violation of his beliefs in all-round education • It is important to sow seeds and to have something rooted in the Ss – providing Ss with "integrated learning experiences" can train Ss to "look at things from different perspectives" and make them aware of "the existence of certain resources" that are available for lifelong education	**Dynamic / theory-laden view:** • Science is the *conceptual re-ordering* of nature. • Using figures and symbols to simplify abstract things is an ever-lasting theme of science • Every person is a scientist in a way. There is theory making in serious science. It can also make use of abstract theory to predict other situations or to bring in *new* experiments. • Science can have many different sources of origin. • Science education is to develop citizens who can actively solve their problems	**Affective / cognitive / motor skills:** • Acquire manipulative skills • Consolidate theory learning • Extend Ss' knowledge scope • Learn different ways to retrieve information (broadening the definition of practical from hands-on activities to include multimedia, simulations, etc.) • Realization of the concept of unity and diversity – putting seemingly unrelated things together through biological knowledge. • Cultivate curiosity	**Learning-focused:** • Teacher's role is to create "invitations to learning" which once picked up by Ss will help them to learn further on their own • Ss' motivation is a prerequisite for learning • Learning takes place not only in school but everywhere • Real learning requires in Ss a systematic structure to organize the information which has to be "memorized" • Teaching should emphasize the thinking process and make use of the teaching situation to guide Ss to elevate step by step
Dawn	• Aims at help Ss get through the exams • Regards cultivating Ss' interests in biology as a bonus • 3 mottoes guiding her teaching: Biology is fun. Biology is part of everyday life. Biology is not difficult.	**Static / objective view:** • Science is to discover. • Science is a method but it does not necessarily involve experimentation. • Science has its own limitations. • Science learning is learning for life because of its *objective* and reliable method, which can be made use of to master other kinds of learning.	**Affective / cognitive / motor skills:** • Stimulate Ss' interest by giving them a sense of wonder • Consolidate theory learning by making things concrete and visible • Learn scientific method and acquire a scientific attitude	**Learning-focused:** • Teacher is a facilitator – to create in Ss "an inner desire, a passion to learn" • Teaching without learning is nothing. Learning is the duty of the student. Ss have to play an active role • Allow Ss to work on their own experimental design – ownership is important to Ss' learning • Knowledge must not be put to Ss in an "elite" way • Peer learning pedagogy is important

Table 12-1. Continued

Teacher	Personal experience/vision /aspirations	Beliefs about science and science teaching	Beliefs about role of practical work	Beliefs about teaching and learning
Eddy	• Teaching is drilling Ss for exams in order to maintain a high passing percentage to satisfy the principal. • I was a "money hunter". • I am a person with no vision.	**Static / objective view:** • Science is *logical* and *systematic* and is not produced by trial and error. • Science is supported by evidence which is repeatable and testifiable. However, "some are still theories" which cannot be verified through experiments. • Observation is the most basic thing in science.	**Little thought about it:** • Didn't give detailed thoughts about this; work has been mainly governed by the syllabus. • Upon further probing, assign priorities to acquiring of basic techniques, cultivating a scientific attitude and developing interest in the subject.	**Teaching-focused:** • Teacher must "build up a strong sense of authority" and a "well-prepared" image in front of the class in order to have more efficient teaching and learning. • A teacher has "to feed" Ss a lot. • Inductive approach of teaching is time-consuming and difficult to "control". • Ss like teachers who can give them a "rich body of knowledge, especially those of the exam techniques."
Fay	• Learning things is not for something special but just to enrich oneself. • Would not limit her conversation with Ss to biology only • Treasure every single student as a human who has his own potential	**Static / objective view:** • Science is a study of the theories governing the earth and living things • Scientific theories *remain unchanged throughout time* • Science is very *objective*, based on theory, and based on evidence; it has predictive capability, and is about logical thinking but it is not "imagining something out of a vacuum"	**Affective / cognitive / motor skills:** • Master basic lab skills • Practise problem solving • Investigate Ss' own problems (but not practised because of the perceived time constraint)	**Learning-focused:** • Teacher must have a strong knowledge base in order to provide Ss a broader perspective to look at things. • Teaching must involve systematic, intentional integration of information in a way such that the recipient can benefit from it. • The key of learning is with the learners, their attitude and their desire to acquire knowledge.
Glen	• Cares about Ss' academic performance and puts pressure on them in this respect • Prepares Ss for tertiary studies • Summer vacation better reserved for revision than for carrying out individual projects • Begins to realize the importance of developing independent thinking in Ss	**Static / objective view:** • Science is to explain things in nature and to find out the *truth*. • Asking questions is a very important step in the process of doing science. • High level science involves a discussion of how evidence can support theories. • Thinking is a part of science.	**Little thought about it:** • Practise problem solving which would build up Ss' confidence • Verify theories taught • Develop basic lab skills • Develop interest • An adjunct to theory teaching • But confessed that not much thought had been given to this	**Teaching-focused:** • Teacher as 'parent' of Ss, providing them with all what they want • Peer learning pedagogy is no good. For teaching to be effective, there must be a teacher.

Table 12-1. Continued

Teacher	Personal experience/vision /aspirations	Beliefs about science and science teaching	Beliefs about role of practical work	Beliefs about teaching and learning
Hugo	• School education is to prepare Ss for their future working life • Education is to cultivate in Ss an attitude to live a more optimistic way of life and to look at things from a wider perspective • Cultivating in Ss a proper value system is more practical and useful compared with imparting book knowledge to them.	**Static / objective view:** • Science is an *objective* method which can be used to solve problems, but there is no exact procedure to follow. • Simply delivery of facts is not science. • Teaching science is letting Ss know the possible ways to find out the answers.	**Instrumental goals outside PW:** • Roles transformed to achieving goals outside the remit of science education *per se*: that is, learning how to cope with problems and to find out one's own strengths • Develop Ss' interest, confidence and sense of achievement	**Teaching-focused:** • There must be teaching taking place in a demonstration • Ss' less than satisfactory performance is due their low abilities, lack of initiative and poor training in earlier years
Ivor	• A knowledgeable teacher is one who knows the syllabus and the trends of exams very well. • Ss will not learn if they are not serious. Assessment makes Ss more serious about learning. • Biology knowledge is disjointed parts, un-related to each other.	**Amorphous view:** • Textbook definitions of science: a systematic study which requires people to think in a scientific way. A complete scientific method should comprise observation, hypothesis, test, …. etc. • Science is to have a suspicious attitude and to ask as many questions as possible.	**Instrumental goals outside PW:** • Ideally, this is to train Ss on the scientific method and to arouse their interest • In reality, to get a better assessment mark in the TAS	**Teaching-focused:** • The most important thing for a teacher is a good teacher-student relationship. • Ss should ask the teacher and not their classmate. Learning is mostly guided by the teacher. Learning alone won't achieve much. The progress of the lesson is not controlled by the student but by the teacher. • Ss can learn better when provided with good and concise notes from the teacher
John	• My work is always driven by exams.	**Amorphous view:** • Science is difficult to define. • It has its own scope… defined over the years… • Carrying out an experiment by following a manual is science, too.	**Instrumental goals outside PW:** • Satisfy the requirements of the TAS to differentiate Ss • Prepare Ss for the theory exam because: • Practicals supplement the theory taught • Report writing and graphic skills also help Ss in the theory exam.	**Teaching-focused:** • Conveying information is teaching. • Some teaching can be unidirectional • Teacher's role is to highlight the key points, teach Ss how to study…

Such a handy reference is deemed necessary and important for readers to judge if the interpretations I made of a particular teacher at different places in the book are trustworthy by cross checking with information contained in this table. Indeed, as pointed out by Wallace and Louden (2000) and evident from the case stories, the meaning of teachers' actions only becomes clear when viewed in the context of their personal and professional histories, their hopes and dreams for teaching.

As evident from Table 12-1, different teachers have different visions of education. These varying visions of education, in turn, rest on ideas about knowledge itself, how learners come to have it, and what knowledge is of greatest value for the future for which learners are being prepared. Thus, it is not surprising to find that teachers who tended to focus on preparing students for the examinations were, in general, more teaching-focused (e.g. Eddy, Ivor and John) than learning-focused. These teachers tended to see their roles mainly as dispensers of knowledge, transmitting information to the students, directing student actions and emphasizing static knowledge. On the other hand, the teachers who were more learning-focused (e.g. Alan, Bob, Carl, Dawn and Fay) tended to view their roles as facilitators to help students process information, coaching them, facilitating their thinking, modeling the learning process and using teaching materials to cater for individual learning needs. They were also the teachers who tended to look to goals beyond that of preparing students for the examinations. There are, however, aspects of the teachers and their beliefs that are less easy to classify in this way. This is particularly so in the cases of Hugo and Glen.

With respect to their beliefs about the nature of science and the role of practical work in school science, there were also wide variations among the ten teachers. On the one hand, some teachers saw science as dynamic and on-going and recognized that observation is theory-laden (e.g. Alan, Bob and Carl). On the other hand, some teachers saw science as objective and/or considered that the knowledge it produces remains unchanged throughout time (e.g. Dawn, Eddy, Fay, Glen and Hugo). Yet, there were others who had only vague ideas about the nature of science, or were unable to articulate them (e.g. Ivor and John).

Regarding their beliefs about the role of practical work, some teachers focused on developing appropriate scientific attitudes in students, the kind of cognitive skills and motor skills required to carry out scientific investigations or problem solving (e.g. Alan, Bob, Carl, Dawn and Fay). However, other teachers admitted that they had actually given little thought to the role of practical work in school science (e.g. Glen and Eddy). Yet, in other cases, the teacher seemed to have transformed the goals of practical work teaching towards achieving some other instrumental goals such as preparing students for their future working lives (e.g. Hugo), or preparing them for public examinations (e.g. Ivor and John). In sum, there was quite a range of beliefs and aspirations held by the group of teachers studied in this project.

2. TEACHERS' ENACTMENT OF THE TAS

Not surprisingly, given the wide range of aspirations and beliefs held by the teachers, there was enormous variation among the teachers in their conduct of the TAS practicals and interpretation of the TAS regulations. Within the case stories, there is a strong suggestion that TAS practicals which were conducted with a group of students enjoying their teacher's prompting and guidance throughout, elicited the students' best performance, and avoided most anxiety – as in the case of Alan, Bob, Carl and Dawn. At the other extreme, where the teacher offered little or no help in the process, the students were likely to perform well below their optimum level and their performance would be likely affected by anxiety – as in the case of Hugo, Ivor and John.

Thus, the evidence reported in the case stories lends support to Broadfoot's (1995, p.34) concern about the possible "grave injustices" that may be done to many schools and students in a high-stakes assessment system, because of the mediatory effects of teachers' varying professional strategies. Such large variations in the teachers' assessment practices call into doubt the validity and reliability of the TAS. This issue is not, however, the main focus of this book. Rather, my concerns centre on those changes in normal classroom interactions brought about by implementation of the TAS, especially those where teachers refrained from helping their students in order to maintain fairness of the assessment. What were the teachers' beliefs that underpin such classroom actions? Evidence presented in the case stories collectively and convincingly point to the fact that there was a close relationship between the teachers' classroom practices and their perceptions and/or understanding of the TAS. That is, whether the TAS was perceived primarily as an assessment reform or one with a pedagogical reform dimension, too. In other words, the classroom practices of the teachers were largely a consequence of the interactions between the teachers' perceptions and/or understanding of the TAS and their various other beliefs, amongst which were their beliefs about the nature of science, the role of practical work, their beliefs about teaching and learning, their visions about the goals of education and their personal aspirations.

3. TEACHERS' UNDERSTANDING / PERCEPTIONS OF THE TAS

In order to facilitate moderation of marks across different schools (which might have adopted different marking standards), HKEA requires teachers to rank order their students when submitting the mark sheets at the end of the A-level course. Such an administrative requirement, coupled with teachers' general perception that the public examination was a norm-referenced assessment, led many teachers in this study to believe that their main role in TAS was to differentiate students' practical competence "accurately" and "to very minute differences".

Not only did such a belief affect the teachers' planning of their lessons, but also their interaction with the students during the TAS practicals. For example, some teachers like Hugo and John thought that it was essential to avoid interacting with students or offering assistance to students during the practical, otherwise students were going "to get very similar marks". They would then be unable to achieve the goal of differentiating students' performance. In one extreme case, his perception of the need to differentiate students had forced the teacher, Ivor, to adopt a "picky and fault-finding" attitude when assessing students even though he knew that it was unfair to his students. On the other hand, teachers such as Carl thought that, "it won't be a problem even if all students are performing well in a particular assessment". According to these teachers, as Bob maintained, "the spirit of the TAS does not necessarily require teachers to differentiate their students very accurately". Hence they aimed at ranking students' performance in very broad terms only–"bright, mediocre and less bright". Thus, based on the interview data, teachers in this study can be separated into two groups, according to whether or not they perceived precise differentiation of students' performance as their prime goal in the TAS, as shown in Table 12-2.

On the other hand, based on their classroom practice, teachers can be rank ordered according to their increasing concern with assessment-related matters (in terms of % DTUs) during the TAS practicals (see Figure 3-3). Table 12-3 shows the ten teachers in order of their increasing concern with assessment-related matters during the TAS practicals. The close resemblance of the *grouping* (not rank order) of the teachers in Tables 12-2 and 12-3 suggests that those who perceived their main role in TAS as being to differentiate students as precisely as possible were also those who tended to be more concerned with assessment-related issues. Indeed, the evidence presented in the case stories corroborates this conclusion.

Table 12-2. Teachers' perception of their main role in TAS

Goal seen as precise differentiation of students' practical competence	
No	Yes
Alan, Bob, Carl, Dawn, Glen	Eddy, Fay, Hugo, Ivor, John

Table 12-3. Rank order of the teachers according to their concern with assessment-related issues

Teachers arranged in order of their increasing concern with assessment-related issues →									
Alan	Bob	Carl	Dawn	Eddy	Fay	Glen	Hugo	Ivor	John

4. TEACHERS' PERCEPTIONS OF THEIR ROLES AS ASSESSORS

Before moving on, it is important to recognize the emotional turmoil evident in many teachers' accounts of their current practice in the TAS. The case stories collectively illustrate the kind of pressures and intense anxiety experienced by many teachers as a result of the implementation of the TAS. These have arisen mainly because of the dual roles of teachers and assessors which the teachers were required to act out in the TAS. These dual roles were necessarily influenced by the relationship of the teachers both with their students and with the public examining body. As teachers of individual students, they were clearly involved in a pedagogical relationship, concerned with furthering their students' learning in biology. They also wanted to ensure that each student could achieve as high marks as possible in the external examination. At the same time, however, in carrying out the TAS assessment the teachers were also acting as agents of the external examining body and, as such, were guided by the TAS regulations set down by the HKEA. If they failed to abide by those regulations, and the moderation process resulted in major changes in students' grades, they would probably lose their standing in relation to students, parents and the school.

The complexity of performing the roles of both teachers and assessors meant that there were contradictory value systems at work, giving rise to tensions or pressures as experienced by the teachers concerned. This, in turn, affected the teachers' classroom actions in one way or another. Thus, in some cases (e.g. Eddy, Hugo and Ivor), the teachers regarded the TAS as an activity distinct from 'teaching', and one which they were being asked to conduct in order to gather evidence for a third party – for assessment purposes – rather than to benefit their students. Indeed, how teachers perceived their roles as assessors was one of the determining factors affecting their classroom actions during the TAS practicals. Interestingly, four metaphors were used by different teachers to describe their roles as assessors in the context of the TAS:

- Alan saw himself "not as an adjudicator or a spokesman of the public examining body but as a *students' companion* who is with them to cross over the hurdle of the TAS"
- Eddy saw himself as "the *referee of a ball game*"
- Hugo saw himself as "an *examiner of a driving test*"
- Ivor saw himself as "a *policeman who fears to be scolded by his superior* for not being able to give out the assigned quota of illegal parking tickets"

As evident from the case stories and the metaphors they used to describe their roles as assessors, it was very clear that Eddy, Hugo and Ivor adopted an authoritarian stance in dealing with disputes with students over deduction of assessment marks. On the other hand, Alan held quite a different view of the same issue. He wanted to gain students' trust of him as a fair assessor through his "gestures and attitude", rather than through the teacher's authority. As he put it, "I would tell them through my attitude and gestures that I won't be prejudiced against a particular student. This would then cultivate in them a confidence in me."

These metaphors reveal somewhat different bases upon which the student-teacher relationships were built. At one extreme, there were teachers, like Eddy, Hugo and Ivor, who took on an authoritarian stance in which power is always in the teacher's hand and the students are always subordinate. It was this belief that gave teachers confidence in dealing with the issue of fairness when disputes arose because it allows the teachers to preserve their power position. At the other end, although teachers are always in a particular power position relative to students, some teachers, like Alan, had been working towards the establishment of "power with" (Kreisberg, 1992, p.61) in their class. That is, the power which is potentially empowering for all members of the class. In the context of investigating classroom-based formative assessment, Torrance and Pryor (1998) suggested that fruitful ways of working towards recursively creating and utilizing "power with" include paying attention to issues of language structure, motivation and the appropriation of joint knowledge through classroom interaction. Indeed, much of these were evident in Alan's classroom practice (See, for example, Episodes 4-3, 4-5, 4-6, 4-7).

Overall, the above findings lend support to Tobin and Fraser's finding (1989) that the metaphors with which teachers conceptualize their roles affect their teaching practice in the classroom. They also highlight the need to prepare teachers for the new assessment era. For teachers to implement the new scheme successfully, their existing understanding and beliefs concerning assessment must be challenged and opportunities provided for them to come to terms with the philosophy of the new assessment scheme. One way is to help teachers to reflect on the metaphors they use to describe their work. This is especially true in cases where metaphors are spontaneously revealed, as in this study, unlike those in cases where metaphors are deliberately used as an instrument to tap into the tacit referential systems that govern teachers' actions, as in much research. This is because teachers under such circumstances might not be aware of the associated tacit beliefs at the time they articulate the metaphors. In that case, the teachers need assistance in making their tacit beliefs explicit, and to objectify them before they can be further reflected upon. Elsewhere, I have proposed a three-step professional development model to help teachers to become metacognitively more aware of their own beliefs and assessment practices through the use of metaphors (Yung, 2001).

5. EXTENDING THE ANALYSIS BEYOND THE PEDAGOGY-ASSESSMENT DICHOTOMY

In looking across all the cases, three recurring themes have emerged. The first is the teachers' concern with fairness, be they assessment-oriented or not. For example, Ivor and Alan adopted quite different measures in terms of giving the same time allowance for all students to complete their work (Ivor drew a red line on students' report when the time was up – see Episode 10-6; Alan required all pupils to hand in their reports within 30 minutes after the first pupil had handed in the report – see Episode 4-6), yet they both mentioned the notion "fairness" as guiding their actions. It is beyond doubt that both Alan and Ivor meant what they said. That is, they really thought that their actions were fair to their students. But why were their actions so different? What did they actually mean when they said "fair" or "fairness"? And how was the notion of fairness related to their classroom actions?

The second recurrent theme that has emerged in the course of re-telling these case stories is how differently the teachers interpreted the same TAS regulations. For example, Ivor thought that teachers should not discuss things with individual students, otherwise it would constitute cheating, while Carl did not think so. The question arises: *Why* did the teachers interpret the TAS regulations differently?

The third recurrent theme emerging from the teachers' voices is the teachers' sense of professionalism. In general, the more the teachers saw the TAS as a pedagogical reform, the more they experienced a sense of professionalism in their work. Table 12-4 summarizes some of the teachers' perceptions of how the TAS had affected their professional development. As revealed by the range of expressions like "The TAS *requires* us to do it this way" to "The TAS is the *driving force*" and "The TAS has *stimulated* me …", the effect of the TAS on the teachers' professional development had not been so uniformly positive as might have been anticipated (e.g. Pang, 1992). Teachers themselves had to notice opportunities for learning and professional development, or stimuli for change in their immediate environment. However, such noticing was not automatic. Some teachers noticed many opportunities for learning while others noticed few or none at all. In other words, professional development was largely an outcome of the teachers' conscientious and informal engagement with the ramifications of the TAS, as well as their sense of professionalism. Thus, a fruitful way forward would be to explore the second and third recurrent themes together, rather than separately. That is, to address the following two questions side by side: Why did the teachers interpret the TAS regulations differently? Was this related to their sense of professionalism? Indeed, closer reading of the interview transcripts along the three themes identified above has enabled me to extend the analysis of the data beyond the pedagogy-assessment level of interpretation, as I report in the next two chapters.

Table 12-4. Teachers' perception of TAS as affecting their professional development

Teacher	Statements from interview excerpts	Inference on how teachers saw their role in the TAS reform
Alan	The TAS … has catalyzed this to occur… The TAS is the driving force… Teaching is a long process of struggling… My feeling is that I am improving year after year.	Teacher as the 'creator' of his own teaching with the TAS as a driving force for him to excel in his teaching.
Bob	I have to take a lot more into consideration in my planning… I have to think over it carefully… This is a good influence on my teaching.	Teacher as the 'creator' of his own teaching with the TAS exerting a good influence on his teaching.
Dawn	Now, I realize that it's good if there can be room for students to develop variations amongst themselves… I share the same view (as that of the TAS) that, in the long run, students should acquire this ability... Students enjoy doing it this way. Their feedback drives me to think about this further…	Teacher as the 'creator' of her own teaching with the TAS providing opportunities for her to try out the new teaching approach which she found students enjoyed and learned well from.
Hugo	The TAS has stimulated me to think retrospectively about what exactly is the biology practical?… I have benefitted a lot from it.	Teacher as a 'reflective practitioner' with the TAS as a stimulus for him to reflect on his own practice.
Ivor	I had no idea at all how to teach these… I had no confidence at all. The resources produced by the TAS … are of great help… I have more confidence now.	Teacher as a 'technician' carrying out someone else's plan – with the TAS resources guiding his teaching, which in the end has helped build up his confidence in teaching new curriculum concepts.
John	I created a scenario… The TAS *requires* us to do it this way… This is the trend.	Teacher as a 'technician' following someone else's instruction.

Chapter 13

THREE VIEWS OF FAIRNESS

A close examination of the cases revealed the existence of several different belief systems. In order to develop an understanding of the relationship of these different belief systems prior to discussion of the relationship between the teachers' beliefs and their classroom actions, I consulted Pajares's (1992) very thorough review of "*Teachers' Beliefs and Educational Research*", in which he identified 16 characteristics of beliefs as they relate to teachers and educational research. Of import to this study are the notions that a person does have several systems of belief; these can be prioritized; the earlier the establishment of beliefs, the more they are resistant to change; and, perhaps most useful to the present analysis,

> Belief substructures, such as educational beliefs, must be understood in terms of their connections not only to each other but also to other, perhaps more central beliefs in the system. (Pajares, 1992, p.325)

This insight has been most helpful in interpreting what I saw as developing connections between the teachers' beliefs about science, teaching and learning. Some of these central beliefs were associated with the teacher's sense of fairness (discussed in this chapter) and their sense of professionalism (discussed in Chapter 14). As the data show, these beliefs, in turn, interacted with the other beliefs to determine the goals which the teacher wanted to pursue and, hence, their classroom actions. This explains why studies attempting to link teachers' beliefs to their classroom practice have sometimes yielded confusing results: one possible explanation is the researchers' failure to identify the "more central beliefs" in the belief system of the teacher.

As evident from the cases, one theme emerging from the data was the teachers' concern with the notion of fairness. All the teachers in the current

B.H.W. Yung, (ed.), Assessment Reform in Science, 165-182.
© 2006 *Springer. Printed in the Netherlands.*

study had been trying to make their classroom practices as fair as possible for all their students while carrying out the assessment. They were all struggling with the issue of fairness in implementing the TAS, but in three qualitatively different ways, as exemplified through the case extracts reported below.

There are several reasons why extracts from just three cases are reported here. First, there is limited space available in this book. Second, these three cases cover all the different views of fairness identified amongst the teachers. Third, the purpose of this and the succeeding chapter is to identify the major themes of teachers' practice that are the most critical for further analysis. The major themes so identified provide a level of commensurability for looking across all the cases.

1. JOHN : I MUST BE FAIR

To recapitulate, John is at the extreme right hand side of Figure 3-3. That is, he was the most assessment-oriented of the teachers. As revealed in Chapter 11, John's practices in his teaching of practical work were closely related to his belief in the importance of preparing his students for getting good examination results. He informed students in advance if a particular practical was going to be an assessment practical. That is, special assessment sessions were set, rather than such skills being assessed in the course of ordinary practicals as originally intended by the TAS. Such an arrangement made the situation very much like sitting for an examination, and created pressure on the students, who then requested John to allow them to enter the laboratory earlier so that they could psychologically prepare for the assessment.

Episode 13-1 – Coming into the laboratory earlier
At the students' request, John allowed his students to come into the laboratory 15 minutes before the practical was officially due to start.

The rationale behind John's decision to allow students to come in earlier was:

> They [the students] take assessment very seriously. They want to have a look at the apparatus first and sort of prepare themselves psychologically for the practical. So, I allow them to come in earlier during the lunch hours... Some of them have very high expectations of themselves... They behave this way whenever marks are counted, be it theory or practical assessment. They have been brought up this way.

Students' obsessive concern with assessment was a prominent feature in John's classroom, as in many other classrooms in Hong Kong (e.g. Pong and Chow, 2002). This is a reflection of their culture, as John put it – "they have been brought up this way". The act of allowing students to come into the laboratory earlier was heavily laden with assumptions rooted in cultural myths – "taking assessment seriously ... high expectations ... whenever marks are counted," etc. Thus, instead of alleviating the pressure on students which is associated with the one-off external practical examination, the TAS – in the way it was implemented by John – had in effect aggravated the situation by exerting a lot more pressure on students throughout the two-year A-level course. As John put it, "The TAS has driven my students to 'death' already". Though John blamed the culture, he seemed unaware that he himself was perpetuating this culture, as can be seen in the following episode.

Episode 13-2 – Free of charge

John started the lesson by distributing the lab manual to the class and allowed them some time to read it. He then invited questions:

TWC: This is a Type I experiment [where students are expected to be given very limited guidance and to work on their own] [...] Any questions before we start? [...] Any questions, please?

There were no questions from the students. John then said again:

TWC: Come on, any question? Free of charge! Marks will not be deducted. Come on. Any question?

Again, there was no question from the students. So John signalled the class to begin their work.

The real meaning of this episode lies in the shared meaning it has for the participants and not in the particular words uttered. As suggested by Bell and Cowie (2001), assessment takes place in the social space of the classroom. It is a social practice, constructed within the social and cultural norms of the classroom. It is shared; it is an object of awareness. It is not just a mental thought object residing inside the head of the teacher. Why were there no questions from the students? The amount of help provided to students may have constituted one of John's criteria for assessing his students' practical competence. Such an assessment criterion had been reiterated time and again by John during the introductory session to his class about the TAS. Thus, John's stating that "marks will *not* be deducted" at the beginning of the lesson might just have reinforced students' perception that "marks *will* be deducted" in other situations where the teacher does not invite questions from them. As will be evident in subsequent episodes, students in general did not like this idea. Often they preferred to proceed without assistance even though they realized that they might not be able to generate an effective response to the practical task assigned.

In fact, it was very rare for John to invite questions from the class during the TAS practicals. This occurred only at the beginning of the practicals observed, where the intent was mainly to sort out problems related to provision of apparatus and materials. In the actual course of the TAS practicals, John was reluctant to answer students' questions, as illustrated by the following episode.

Episode 13-3 – I have a question but will marks be deducted?

In the middle of the practical, a student wanted to ask John a question. The dialogue was as follows:

S: I have a question but will marks be deducted?
T: You ask it first.
S: Chee! I don't want to ask then.
T: If I am going to deduct marks, I will tell you first.
S: If I ask you the question, but then you tell me afterwards that marks have been deducted, I will be very depressed.
T: Just go ahead and ask me. And you will know what the outcome would be.
The student then asked the question.
T: I have to deduct marks from you if I answer you this question. Therefore, I am not going to answer this question. You think about it yourself.
S: Are you really not going to deduct any marks from me at all?
T: Go back and do your work quickly.

The reason behind John's decision not to answer the student's question was that:

> I must be fair. I can't answer some students' questions but not the others. If I answer her question, I am sure that I will have to tell the rest of the class using the microphone... What bothers me is that, suppose I am going to answer students' question, how many questions should I entertain, and to what extent? This is the most difficult part. If there was no TAS, I would then give her a definite answer ...

John was caught in a dilemma of trying to be fair to all students on one hand and trying to help solve their problems on the other. Obviously, teaching and assessment had become polarized; teaching had given way to assessment, and the formative function of assessment was lost. John's perception of assessment as purely for summative purposes tempted him to undertake assessment in formal testing situations where students' requests for help were not entertained. In the long run, such a practice could have an adverse influence on the student-teacher relationship, as implied in the following two episodes.

Episode 13-4 – You go away because you didn't answer my question
John was standing next to a student watching her doing her work, the student said to him, though in a joking manner, "You go away because you didn't answer my question."

When asked to express his feelings on this episode, John said,

> She won't blame me just because I didn't answer her question. I think that she just did not want me to disturb her.

One of the important external influences on learning is the classroom environment, and this includes the student-teacher relationship, and the teacher's style and competence. These determine the overall quality of the classroom climate the teacher creates and the quality of his interaction with individual students. In the above instance, John might not be aware of the possible damaging effect on the student-teacher relationship of not entertaining students' questions. The possible deterioration in the student-teacher relationship in this classroom is revealed in the following episode, where it took John some effort (asking a student four times before getting a positive response) to determine what sort of difficulties the student was encountering.

Episode 13-5 – No, no problem. No problem.
John noticed that two students were discussing. He had a feeling that they might be encountering some difficulties with the practical. Thus John intervened and asked one of the girls:
T: Is it that you don't know how to do it?
S: No, no problem. No problem.
T: Any problem?
S: No problem. No problem.
T: Just ask me if you've got any problem.
S: No problem.
T: Is there any problem? You don't know how to do it? Any problem?
S: I forget what is the meaning of 'hypotonic'.
T: Hypo means a greater water potential.

When asked about his feelings about this instance, John expressed his feeling of being dis-empowered by the TAS, though rather implicitly:

> Students generally do not want to reveal their difficulties or weakness to the teacher. Maybe their impression is that marks are going to be deducted... I just don't know why she did not want to ask me. What can I do?

Obviously, the effect of the TAS on the teachers' professional situation was not so uniformly positive as anticipated. This episode illustrates how

John's professional situation was undermined. Before the implementation of the TAS, teachers were to a large extent free to run their classes and laboratories as they saw fit. There were accepted canons of practice and a body of "teacherly" knowledge about how things were done. The latter is a form of tacit and practical knowledge which is sustained through traditions (Nyiri, 1988). However, the TAS requires teachers to construct a novel professional practice which entails a shift in their understanding of the role of assessment, as Gipps (1994) suggested, to an "educational assessment paradigm" rather than a traditional "testing paradigm". However, John's adherence to the traditional "testing paradigm", as evident in the episode below, made him felt somewhat dis-empowered.

Episode 13-6 – Fearing that none of the students would be able to accomplish the task
John found that most students were stuck at one point during the practical. He was talking to himself, fearing that none of the students would be able to accomplish the task. Instead of circulating and supervising students' work, he went back to the teacher's bench and sat there with his hands supporting his head, engaged in deep thought for at least two minutes before announcing to the class a crucial step in carrying out the practical.

In the post-lesson interview, John admitted that he was struggling about whether he should give the hint to the class or not. He explained:

> I had never thought of telling them everything [about the procedure of the experiment]. Given the abilities of my students, if you tell them the crucial point, you are just making trouble for yourself. It will be difficult for me to differentiate them [that is, to rank order their performance]...

1.1 What do the episodes tell us?

The above episodes, together with those presented in Chapter 11, make up an instructive case story of how firmly held ideas about assessment can affect the quality of teaching and learning at the classroom level. John's first and foremost considerations were differentiation of the students' capabilities and the notion of fairness, rather than the wider concern with how assessment could best be used to promote learning. The particular tensions and pressures perceived by John are further revealed in this interview excerpt.

> The TAS is certainly an assessment. I have been criticised by my chemistry colleague that the TAS is training and not an assessment. But, the hard fact is that I have to submit the marks to

> HKEA. It is no good placing too much emphasis on learning ... If you tell students that marks are not important, just to relax and try, students just would not believe in what you say, especially students in this school. They know what is going on. You just can't fool them.

The requirement to submit the TAS marks to the public examination body for certification purposes had framed the way in which John interpreted the assessment reform. He had drawn on his previous experience and understanding of what assessment was about in order to make sense of the changes and to make decisions about how he should implement the TAS. The above episodes highlight vividly how such an interpretative framework was deeply established in John's belief system. He felt under pressure to set aside special assessment occasions that were more controlled than routine practical work sessions and not to answer students' questions during the course of the practicals. This is not surprising if the introduction of school-based assessment is regarded as purely for the purpose of improving the validity and reliability of the assessment. In fact, this was not an unreasonable assumption for teachers to make when the reform was initiated by the HKEA. No wonder that, based on such a mindset, the classroom situation in which practical work occurred was that of a formal assessment. Though this did not mean that the teacher could not help the students, any assistance given would be an integral part of the assessment process, and would result in the deduction of marks. In other words, the teacher was explicitly acting under the authority of the HKEA, as an extension of the examination procedures. This aspect of TAS work was a source of tension for a number of teachers in this study, including John. For these teachers, concerns about how the reform may broaden the curriculum and improve the quality of teaching and learning had drifted quietly into the background.

In summary, this case shows that genuine improvements in the effectiveness of learning actually require a major rethink about the way that assessment is used. This rethink needs to be based on a careful analysis of how assessment can promote individual learning. In particular, in-service training for teachers aimed at re-constructing their ideas of the role of assessment in educational reform and helping them to develop strategies for utilising the outcomes of assessment in formative, evaluative and educative respects are imperative (Hodson, 1992). To this end, the following two cases are illustrative.

2. BOB : THEY ARE LEARNING WHILE I AM ASSESSING THEM

Compared with John, Bob seemed to be able to cope with the requirements of the TAS quite well. He also seemed to be able to come to grips with the formative function of the assessment, often initiated discussions with individual students during the course of the practical. In fact, he was mindful of the importance of interacting with students even at the lesson planning stage:

> One of my major considerations in selecting practicals for the TAS assessment is whether I can make use of the practical to induce some kind of discussion with my students and that they can learn through it, something that they have not thought about before. This is a very crucial part in their learning. That's why I always ask them questions continually throughout the practical. So, they are in fact learning while I am assessing them.

Sadler (1989) believes that assessment is truly formative only when it involves the student. She conceptualises formative assessment as being concerned with how judgments about the quality of students' responses can be used to shape and improve their competence by short-circuiting the randomness and inefficiency of trial-and-error learning. This is exactly what Bob was trying to achieve when he engaged in active discussions with his students, as revealed in the following two episodes.

Episode 13-7 – You think it over yourself first. I'll come back to you later.
T: I have a question for you. Why have you put the epidermal peel on the white tile?
S: I've just torn it off from the leaf.
T: Have you thought of putting it in water?
S: Um! Would it ...?
T: You think that will affect its osmotic potential?
S: No, perhaps according to the theory…
T: So, why don't you put it in water?
S: [...]
T: My question is: Why is it that putting the epidermal peel in water won't affect its osmotic potential? How does it differ from the last experiment we did on potato, where you couldn't put the potato cylinders in water? Put the epidermal peel in water first, if you want to. You think it over yourself first. I'll come back to you later.

Episode 13-8 – Any other better procedures that you can think of?
T: Edmund, hold on. What do you think you are doing here?
S: I leave them [the potato strips] here to dry up first. Shouldn't it be like this?
T: But...
S: Do you mean that this will also affect its water potential?

T: So you know what the problem is. So these two [potato strips] have to be thrown away.
T: So, just as you have said, if you expose the potato strips in air, it also affects the water potential. So what are you going to do?
S: Put them back into the sucrose solutions once I have finished weighing them.
T: Maybe. Any other precautions or any other better procedures that you can think of? (No response from the student for more than one minute)
T: In fact if you say that the faster the better, you try to reduce the time that the potato is exposed to air. So, which step do you need to do first?

So, one common feature in the discussions of Bob with his students was that he always responded to his students' queries with remarks and questions like: "What do you think?" "What better procedure can you think of?" "You think it over yourself first. I will come back to you later." In a post-lesson interview, Bob told me the positive effect of the TAS on his teaching and the learning of his students.

> In the past, I would point out their mistakes directly to them. Now, I have to remind myself to be conscious of this. Telling them directly is the fastest and simplest way, but it does not make them think. This is a good influence on both teaching and learning.

This indicates that Bob was beginning to realise the importance of not only providing feedback to students, but also attending to the quality of the feedback, as Sadler (1998) has pointed out,

> Formative assessment does make a difference, and it is quality, not just quantity, of feedback that merits our closest attention. By quality of feedback, we now realise we have to understand not just the technical structure of the feedback (such as its accuracy, comprehensiveness and appropriateness), but also its accessibility to the learner (as a communication), its catalytic and coaching value, and its ability to inspire confidence and hope. (p.84)

Thus, Bob was able to find a handle or frame of reference outside the concrete situation of assessing his students by being "conscious of not telling students the answers directly so as to make them think". He saw this as a good influence on both his teaching and his students' learning. Marton (1994) describes such a transcendence of one's taken-for-granted experiential world as the ascent to a kind of analytic awareness: a capability of abstracting aspects of concrete situations and seeing these aspects of concrete situations in relation to each other. It seemed that it was the lack of this analytic awareness in John that made him unable to step outside his accustomed frame of understanding assessment for the TAS as a testing paradigm.

Bob felt at ease in discussing with his students on an individual basis, because he felt that he was in control of the agenda and that he was able to know what was going on during the discussion. He also allowed students to discuss amongst themselves although he was a little hesitant.

> I won't intervene [in students' discussion] unless they have been discussing something for a long time. I think this is okay. This may create a more relaxed atmosphere ... In fact, this is a difficult problem for me. Suppose, if there is no assessment, they will learn more from each other through the discussions.

As implied by this interview excerpt, Bob felt that he might be being unfair to his students when cutting short their discussion, as this would deprive them of the chance to learn more from each other.

Even though Bob did not restrain himself from discussing with students, as John did, he was still conscious of the issue of fairness, as revealed in the following episode where he seemed reluctant to provide further hints upon a student's request. He told the student, "I guided your classmates in the same way as I have done it for you. All of them can do it. What has happened to you?"

In this particular instance, Bob was judging the student's performance on a norm-referenced basis, comparing the student's performance with that of his classmates. According to Harlen and James (1997), formative assessments should always be made in relation to where students are in their learning. That is, formative assessment should be criterion-referenced and student-referenced (or isaptive). This means that assessment of a student's work should take into account the particular context of the student's work and the progress he/she has made over time. In consequence, the judgment of a piece of work, and what is fed back to the student, will depend on the student and not just on the relevant criteria. The justification for this is that the individual circumstances must be taken into account if the assessment is to help learning and to encourage the learner. Thus, in the situation described above, Bob's priority of trying to be fair to all students had made him lose sight of his obligation to construct a theory of effective learning which takes contextual variables into account, including students' personal variables. This is unlike the case reported below where the teacher, Carl, was highly conscious of the effect of such contextual variables on students' learning.

3. CARL : IS IT REALLY FAIR?

In the lessons observed, there were many discussions between Carl and his students, and amongst students themselves. For example:

Episode 13-9 – Discuss with your classmates

During the pre-lab discussion, the teacher, Carl, said to the class:

TWC: Yes, it is more accurate. But why is it more accurate? Even though we are running out of time, I want you at this point, to spend about a minute to discuss with your classmates why using ... is more accurate? ... I will come back to you after one minute.

When asked why he often encouraged students to discuss amongst themselves, and whether this would create a dilemma for him in coping with the requirements of the TAS, Carl replied:

> This is a compromise to students' cultural habits of not wanting to be vocal. They are passive. They are unable to respond promptly. I have to give them time to think, to process and to discuss their ideas so as to build up their confidence ... I am aware of the conflict between teaching and assessment, but there is no such formal statement about the DO's and DONT's in the TAS Handbook. I think limited discussion won't affect their overall performance too much. Too much emphasis on assessment will hinder a lot of ideas flowing out. They have undergone the educational process. Is that really going to affect the fairness of the assessment? ... The interaction amongst themselves and between us is an unexplored treasure. I have been encouraging them to speak up. But this has to be built up slowly step by step ... I have faith in my students …

The following episode illustrates how Carl tried to tap into the "unexplored treasure" of interacting with his students during the post-lab discussion.

Episode 13-10 – Each of you makes one suggestion in turn

T: How many of you have recorded the time you started the experiment? I mean the immersion time.
(A few students put up their hands.)

T: Good. I did not tell you to do that. But you have to report the immersion time in your report. Right?

T: Now, how long should the immersion time be?

S1: Half an hour.

T: Why should it be half an hour? Why not 35 minutes? What is the reason behind this?

S1: It is not too long and not too short.

T: Why don't you say 20 minutes? Is that only a subjective judgment? You don't have any reason behind it, do you?
T: Theoretically speaking, how long should the immersion time be? How long should it be? Ference?
S2: Half a day.
T: Half a day? Some say 30 minutes. Some say half day. How could there be such a big difference?
T: What do we mean by an end-point?
T: Gregory? What do we mean by an end-point? What is the end-point for this experiment?
S3: When we obtain the results.
T: When we obtain the results! (laughing) How are we going to judge whether we have obtained the results or not?
S4: When there is no net movement of water into and out of the cells.
T: Yes, it is the end-point when there is an equilibrium. The potato cell does not gain water or lose water.
T: But how do we know? Theoretically speaking, how do we know that it has reached the equilibrium state? Do you know how?
(No response from the class.)
T: You take it [the potato strips] out. You measure the weight. What would you do next? How can I tell if there would be further changes in its weight?
T: Take it out. Measure the weight. What should we do next?
S5: Put it back into the solution.

The class discussion continued for several minutes on several other points before the teacher rounded up the discussion in the following manner:

T: Okay. I leave the rest [of the discussion points] for you to think about yourself. I will leave you alone without further intervention. But you can ask me questions, if necessary. Basically, you have to work on your own. Okay?

As is revealed from his closing remark, in addition to encouraging students to participate in the class discussion, Carl also encouraged students to ask him questions, if deemed necessary, while they were writing up their reports. This is in stark contrast with John, who had no interaction with his students in the report writing stage. When asked if interacting with students so much would affect the fairness of the assessment, Carl's view was:

> This is what science education is about. The TAS never prohibits teachers from responding to questions raised by students. Students' overall performance will not be affected by just one or two points which they might have discussed with the teacher or their classmates. Differentiation [in their capabilities] will be reflected in their overall performance in the reports ... The idea of the TAS is to integrate assessment with teaching and learning.

Very clearly, Carl's way of implementing the TAS was closely related to his 'scaffolding' view of learning, in which the teacher should try to provide a stimulating environment and guide students towards learning "step by step" (see Chapter 6). In all, Carl seemed to be fairly good at integrating assessment with teaching and learning. Nonetheless, in one of the interviews he also raised a point related to the notion of fairness. He was concerned about the common practice among teachers of extending their practicals beyond the normal school hours. Often, the extension could be more than one hour. Carl felt that this was unfair to the students:

> I fear that students may feel bored when I ask them to stay after school every biology practical. This is impossible when we are emphasizing all-round education. Students are encouraged to participate in more extracurricular activities. It is unfair to them if they are denied these options?

Asking students to stay after school until they had finished writing up their reports was in keeping with the TAS regulation that the teacher has to exercise control and supervision over all work assessed. The public examining body thinks that this will ensure that the work assessed is the students' own and thus plagiarism will be prevented. No one will deny the importance of fairness in a public assessment system, but the key is how to achieve optimum fairness for the purpose of assessment, while at the same time still facilitating teaching and learning. We should look at the problem not only at the level of teaching and learning in individual subjects but also in the larger context of all-round education, as Carl put it. There is no simple answer to this problem. This is yet another area where professional judgment is utlilized during the decision making process. This, in turn, is influenced by the teacher's belief system, as revealed in the explanation put forward by Carl for allowing his students, at times, to finish writing up their reports at home.

> There are two main factors that differentiate the students. First is their own background knowledge and language ability. Second, whether they put efforts into it or not. These are the two overriding factors ... If they are not good in language ability or poor in background knowledge, it is impossible for them to improve very rapidly overnight. You can tell immediately if you interrogate them with a few questions. You may say that I have prejudice against some students.

It is understandable that both the examining body and teachers are concerned with improving the credibility of the assessment by making it as fair as possible for everyone. However, the tension between assessing under standardised conditions and providing flexibility to cope with contextual differences in different classrooms is always there. The important point to emphasize here is that the educational benefits derived from the TAS are at a cost to reliability. This cost was paid once the TAS was implemented, signalling concurrence with Harlen's (1994) view that "assessment in education is inherently inexact and it should be treated as such" (p.12). This is not to deny the importance of reliability (and hence fairness): an unreliable assessment is not only of little use but can be unjust as well. The endeavour to increase reliability is common to all methods of assessment but the context and purpose of assessment will affect the degree of priority it is afforded. The key is how to achieve optimum reliability for assessment purposes while maintaining high validity. That is, assessments that are arising as a natural consequence of teaching and learning, not simply something added onto it. To this end, differences in individual classrooms must be taken into account.

In fact, there are many methods of moderation to address the issues of fairness in school-based assessment arising from variations in the marking standard of teachers in different schools and the conditions under which the assessments are carried out. Detailed discussion of this is outside the scope of this book, but certainly, any form of assessment leading to certification cannot be so low in reliability that its validity is seriously called into question. What is beyond dispute is the need to find ways to maximize the educational benefits from implementing the TAS (as Bob and Carl did), instead of feeling constrained by it in the way John described. Otherwise, we are doing injustice to our students – depriving them of the opportunity to learn the subject matter and to receive an all-round education.

4. SUMMARY AND IMPLICATIONS

Case extracts reported in this chapter have shown how the more central beliefs in these teachers' belief systems, in particular their conceptualization of the notion of fairness, can inhibit or facilitate students' learning during the TAS practicals. They also illustrate vividly the kind of teaching-learning activities occurring inside the school laboratory and the various confounding variables affecting laboratory teaching and its assessment.

4.1 Teachers' awareness of the issue of fairness

According to Marton and Booth (1997), a person's awareness is "the world as experienced by the person" (p.108). That is, awareness is the totality of all experiences of an individual in terms of which a certain phenomenon is understood; an experience being an internal relationship between the person and the world. They also argue that "there is only a limited number of distinctly different ways in which people are capable of experiencing any of the things they meet" (p.206). Moreover, the variation in the different ways that people experience something can be understood in terms of the limited human capacity to discern and to be aware of diverse aspects of situations and phenomena at the same time. In short, a particular way of experiencing something represents a combination of related aspects that are simultaneously present in a person's focal awareness. The phenomenon of interest in this chapter is the notion of fairness as experienced by teachers in the context of the TAS. Although the discourses of all three teachers were dominated by, and their classroom actions strongly influenced by, the notion of fairness, this occurred in three qualitatively different ways:

- as an extension of the public examination procedure (John)
- providing chances for students to learn the subject matter (Bob)
- providing students with an all-round education (Carl)

Given that the teachers had been given little support by way of a model for school-based assessment, it was not surprising to find that they had adopted a range of procedures. Though the teachers were carrying out assessments under the same set of regulations governing the TAS, they did so in terms of goals of assessment, in terms of goals of teaching and learning, and in terms of abilities that the students were supposed to develop. These *specific goals* were not considered in isolation, but were seen in relation to more *general goals* such as those listed above (i.e. as an extension of the public examination procedure for John, providing chances for students to learn the subject matter for Bob, providing students with an all-round education for Carl). These specific goals were also seen in relation to the *ways or means* that teachers regarded as likely to *bring about those more general goals*. For example:

- restraining from giving clues to students to help them solve their problems, not answering students' questions, and not allowing students to discuss among themselves were ways considered by John as fulfilling his goal of acting as an extension of the public examination procedure
- asking students a lot of questions in order to make them think was a method considered by Bob to be effective in providing chances for students to learn the subject matter while he was assessing them
- allowing students to complete the laboratory reports at home so that they would not be deprived of the opportunity to participate in extracurricular activities after school was seen by Carl as contributing to his goal of providing students with an all-round education

Accordingly, dealing with goals of teaching and learning implied – explicitly or implicitly – the action of dealing with relations: relations between specific goals and general goals, relations between goals and means. All these depended on the teacher's intentionality; his directedness, what he was oriented towards and in which way (Marton, 1994). In other words, we teach "who we are" and "what we know".

John's consciousness was directed towards maintaining fairness in the differentiation of students' abilities. Bob, for most of the time, was conscious of making the assessment as truly formative as possible by intentionally involving students in discussions with him. He wanted students to learn while he was assessing them. On the other hand, though Carl was able to integrate assessment with teaching and learning of the subject matter very well, he still found the present assessment practice unfair to the students when he looked at it from the perspective of providing students with an all-round education. Carl was actually discerning relevant aspects in the situation and relating them to each other in functional terms. The relations were, in this case, between school-based assessment and students' learning, and between school-based assessment and all-round education for students. This latter consideration brings out a salient point, which is often overlooked by public examining bodies in the process of striving for a credible school-based assessment. That is, the assessment should not make demands on teachers and students that are incompatible with the context in which learning is exhibited. In short, the assessment should not be too time consuming, artificial or divorced from the normal range of contexts in which the educational achievements of students can be observed. Clearly, the extension of practical sessions beyond normal school hours in all the three cases reported above is a case in point, and has put the validity and the educational desirability of the TAS seriously in doubt.

Using the same interpretive framework described in this chapter and based on the evidence presented in the individual case stories in Chapters 4 to 11, all teachers in the study can be identified as holding one of the three different views of fairness as shown in Table 13-1. While the small sample of teachers does not allow me to speculate about the distribution of the various orientations among biology teachers as a whole (or, indeed, to be confident that there are no other conceptions of fairness), it is clear that differences do occur and that the three cases described above are illustrative. This raises general concerns about how teachers can be helped in improving their awareness of the various views of fairness, and their relationships with the different teaching practices in the TAS.

Table 13-1. Teachers' conceptions of fairness

Conception of fairness in the context of	Teacher
Acting as an extension of the public examination procedure	Fay, Hugo, Ivor, John
Providing chances for students to learn the subject matter	Alan, Bob, Dawn, Eddy, Glen
Providing students with an all-round education	Carl

4.2 Developing teachers' analytic awareness on the issue of fairness

Many of the problems which have been revealed are, in fact, contentions and confusion arising from the formative and summative uses of the same set of assessment evidence. Teachers need to be offered help on school-based assessment, and this needs to be done in a way that disentangles the two different kinds of assessment so as to enable teachers to use assessment in a genuinely formative way that helps students' learning. This would include guidance on types of feedback from teachers necessary to maintain student motivation, as well as on identifying specific aspects of attainment or good performance and what can be done to help further improvement (see, for example, Tunstall and Gipps, 1996; Black and Wiliam, 1998; Black et al., 2003). All these skills and understandings need to be developed within initial teacher training and continuing professional development.

Providing teachers with concrete examples that illustrate how the problems can be tackled would certainly help. However, this alone is not sufficient in preparing teachers for the wide range of possible situations in school-based assessment, where the contextual variables among schools vary so much. Teachers *themselves* (with appropriate professional training) need to analytically discern comparatively abstract, generative aspects of concrete assessment situations and relate them to each other (as the present chapter has attempted to do). According to Marton (1994), this is the way in which means-ends relationships or relationship between what is specific and what is general are established. He claims that for a teacher, "the transcendence of his taken-for-granted experiential world means the ascent of a kind of analytic awareness: a capability of abstracting aspects of concrete situations and seeing these aspects of concrete situations in relation to each other" (p.39). He asserts that such knowledge is theoretical knowledge in the sense that it is generalizable and generative. It is, or it can be decontextualized and made explicit. Analytic awareness is a function of theoretical knowledge, the availability of theoretical knowledge.

In the context of the present study, this theoretical knowledge is an understanding of the formative and summative functions of assessment as viewed from the educational assessment paradigm (Gipps, 1994). In this respect, Harlen and James (1997) have offered many suggestions to help teachers cope with their problems and ease their tension. Their guiding principle (theoretical knowledge – in Marton's term) for teachers is to *use* relevant evidence gathered as part of teaching for formative purposes, but to *review* it, for summative purposes, in relation to the criteria to be used for all students. This means that, in the course of the practical, teachers could still help individual students with their specific learning difficulties as well as review all the evidence, for a summative purpose, in relation to a set of criteria that are common to all students. Indeed, the TAS does not bar teachers from offering help to their students as long as they take into consideration the help offered when they assign the final assessment marks (see Chapter 2).

One final point to make is that John's case points to the need to assist teachers in identifying the referents – such as personal beliefs and myths – that they use to make sense of assessment practices. Tobin and Jakubowski (1990) assert that this is an important first step in beginning the process of change. Indeed, the findings of this study concur with those of Briscoe's study (1993) that the reconstruction process is particularly difficult when the referents a teacher uses to assign meaning to salient assessment roles and practices are cultural myths, or beliefs associated with the myths. Accordingly, the impact of the culture of teaching on the construction and reconstruction of cognitive referents must be considered as teachers are assisted in initiating alternative assessment practices.

5. CONCLUSION

Overall, this chapter indicates how complex the assessment process is in the school laboratory, and points to the need for teachers to be aware of its potential for the improvement of teaching and learning and of its side-effects, especially when it is not properly carried out. There is clearly a need for teacher professional development in this vital aspect, especially in terms of developing teachers' analytic awareness on the issue of fairness (see Chapter 15). Assessment reform should be undertaken with full regard to problems of teacher interpretation and mediation at the classroom level. Special attention has to be paid to making clear to teachers the philosophy and intentions behind the new form of assessment – be it through initial or in-service teacher training. Otherwise, there is a danger of teachers interpreting these changes within a traditional "testing paradigm", as in John's case, and this will ruin the intentions behind these new forms of school-based assessment. This would be grossly unfair to all parties concerned – teachers and students alike.

Chapter 14

TEACHER PROFESSIONALISM AND POLICY INTERPRETATION

As pointed out at the beginning of this book, one of the assumptions underlying centrally directed change is that teachers, where necessary, will be both willing and able to adapt their teaching in appropriate directions. Yet there is considerable evidence to suggest that this is not so, as revealed in the case stories reported so far. The teachers mediated the external pressures upon them through the 'filter' of their own professionalism. But the questions remain: why did teachers perceive the reform differently; on what did the teachers base their interpretation of the TAS regulations and their enactment of these in the classroom; was this related to their sense of professionalism?

Hargreaves (1994) defined teacher professionalism as the capacity to make discretionary judgments in the interests of improved student outcomes and transformative educational change. Accordingly, teacher professionalism implies not only the exercise of expertise knowledge but also an altruistic commitment to constantly reflect on, and improve practice in the interests of students. As such, this requires teachers adopting an emancipatory approach (Proudford, 1998) as regards policy interpretation and implementation.

In an attempt to explain teachers' varying interpretations of the policy texts of a curriculum reform, Bowe and Ball (1992) drew upon the work of Barthes (1976) – in particular, the ideas of 'readerly' policy texts, in which the user has minimal scope for creativity, and 'writerly' texts, in which the reader assumes an interpretative role. They argued that the state of being 'readerly' or 'writerly' is not necessarily inherent in the text, but is dependent upon the interactions between the text and the user: in other words, there is a degree of choice. The teacher may effectively collude in the

B.H.W. Yung, (ed.), Assessment Reform in Science, 183-203.
© 2006 *Springer. Printed in the Netherlands.*

diminution of their professional autonomy by unquestioningly accepting the TAS regulations as 'readerly' texts; or they can resist and attempt to subvert and reinterpret them, that is, be writerly.

Helsby (1995) related this notion of professional interpretation to the notion of professional confidence which, according to her, implies a belief both in one's authority and in one's capacity to make important decisions about the conduct of one's work. Another component of professional confidence is the feeling of coping with the work in hand and of being 'in control'. Helsby also asserted that levels of professional confidence may be associated with teachers' ability to maintain a proactive role in terms of managing the changing demands made upon them. Thus, where confidence is high, teachers are more likely to seek to impose their own professional interpretations of government policy and to balance its demands against other professional priorities or vice versa.

Closely allied to Helsby's idea of "to balance its demands against other professional priorities" is Proudford's (1998) notion of professional consciousness. According to Proudford, professional consciousness entails a capacity to problematise taken for granted assumptions and values which underpin policy response and professional practice. She argued that transformative educational change is unlikely to occur unless teachers subject to scrutiny their 'writerly' response to policy documents. Moreover, their 'writerly' responses have to be underpinned by a defensible philosophy and theory of education, and guided by professional consciousness. Proudford referred to this as the emancipatory approach to educational change, and it embraces three key dimensions: professional confidence, professional interpretation and professional consciousness.

As the data emerged from the present study, it appeared that the varying practices of the teachers and the different ways in which the TAS regulations were interpreted by the teachers could well be explained through the lens of the emancipatory approach, underlying which is the notion of teacher professionalism. Analysing the data in such a way has enabled me to gain a deeper understanding of the relationships among the teachers' beliefs, their classroom actions and their sense of professionalism. Due to limited space, extracts from five representative cases will be reported to illustrate the range of teacher professionalism exhibited by teachers in this study. An overview for all the participating teachers will also be provided at the end of this chapter.

1. IVOR : A POLICEMAN WHO FEARS TO BE SCOLDED BY HIS SUPERIOR

The TAS advocates the importance of the formative function of assessment. Teachers in this study generally recognized the need for, and the importance of, such a function in students' learning. This was mainly manifested through their marking of students' laboratory proposals and reports and the discussion of their mistakes with them afterwards. This was, in part, facilitated by the need to carry out the assessment in a more systematic manner in order to fulfill the requirements of the TAS. This not only permitted teachers to provide specific feedback to individual students but also allowed them to make use of the same information to inform and to improve their own teaching, as Ivor put it:

> With the experience gained over these few years, I find that we as teachers can improve ourselves. For example, when I am marking students' lab proposals, I jot down all their mistakes in a file. This helps me to guide them during the discussion... I can see that students have been improving greatly. This is a result of me jotting down all their common mistakes very carefully and then discussing them with the students. I repeat the same process when I mark their reports... This information is also very useful to my teaching in the years to come. It will give me a better grip of the standard of the students. If I accumulate more and more information of this kind, I will be more and more confident with my teaching.

Thus, in the course of implementing the TAS, Ivor had noticed for himself that there were opportunities for his own learning and professional development. This could be seen largely as an outcome of Ivor's conscientious and informal engagement with the changes imposed by the TAS. However, re-examination of the data using the emancipatory approach to educational changes uncovered a rather disturbing scene. In particular, Ivor's 'readerly' interpretation of the policy text could be inferred from his feeling of being constrained by the TAS regulations in his lesson planning and daily teaching, as reflected in the following interview excerpt.

> We have to follow the TAS regulations closely. The TAS requirements occupy a very important position in my lesson planning process. If I do something wrong, the TAS coordinator

> will report to the exam board. I have to be cautious about this. I have to be very careful not to cross the line.

Ivor's 'readerly' interpretation of the TAS regulations was closely related to his low level of professional confidence, as revealed in the following interview excerpt. For example, he was not sure how much guidance he should be giving to his students on the procedure in the experiment when carrying out an assessment practical.

> I don't know what is the exact requirement of the TAS. I feared that I might be giving too much information to students. I really have no confidence because it is very easy to step over the line. The exam board won't accept the assessed work once I have given students too much information. But how much is too much?

In a similar vein, Ivor felt insecure about the possibility of being charged by the examination board with adopting a lenient marking standard in assessing students' work. In order "to play safe", he had chosen to adopt a fault-finding attitude when carrying out the assessment:

> I had to be very picky and fault finding with students otherwise they might get a very high mark. So high that the exam board would not believe in it. I had to behave like a policeman who had to grasp every chance to give out the assigned quota of illegal parking tickets in order not to be scolded by the superior… It is really unfair to them.

Ivor's use of the metaphor of "a policeman fearful of being scolded by a superior" illustrates vividly how he had submitted passively to the TAS regulations, even though he judged it to be misguided, as reflected in his tone and expressions such as "it is really unfair to them". There was clearly a sense of powerlessness and resignation, which in the end had had an adverse impact on his pattern of classroom interaction, as well as the teacher-student relationship, as revealed in the following interview excerpts (see also Episode 10-7):

> I feared that I would be giving students too much help... I was not confident enough. In addition, I felt that with so many restrictions imposed by the TAS regulations... it left little room for me to discuss with them. Hence I'd rather not discuss things with them

> at all. I do worry that I would cross the line. And this would be regarded as cheating.
>
> Once I notice students commit a mistake, I will tell them immediately on the spot. Because I fear that they would not admit that they have committed such a mistake. They would then argue with me during the post-lab discussion that they have not committed such a mistake.

The last excerpt indicates that Ivor's lack of confidence was not just in dealing with new demands of the TAS but also with basic matters like building a trusting relationship between his students and himself. The root of the problem was that Ivor's consciousness was directed to his own self-interest rather than that of his students:

> I worried a lot about how much I should discuss their experimental proposals with them. If I tell them too much, I may be violating the TAS regulations. So, the best thing is that I tell them nothing, I am sort of trying to protect myself as far as possible. I just don't care whether students understand or not. In future TAS assessments, I will not be using this approach again because it is difficult to find out where the demarcation line is.

Clearly, Ivor was more concerned about himself – "I am sort of trying to protect myself" – than his students' learning – "I just don't care whether students understand or not". This example illustrates how a teacher's level of professional consciousness (a low level in this case) may affect his professional confidence and subsequently his interpretation of the policy texts ('readerly' in this case). Ivor's consciousness of "protecting himself" is further revealed in his response to a question about whether it is important to have a short pre-lab briefing immediately before the practical:

> There are bound to be safety issues in every practical. They must be mentioned. Some are very simple and obvious. But I don't mind. This is to protect myself. Even these are very simple things. Mentioning them will help me to protect myself. Our principal is very concerned about this. There is quite a lot of pressure on me. I would be in great trouble if there were accidents in my practicals.

Teacher professionalism implies not only teaching students the subject matter knowledge, but also an altruistic concern for their all-round welfare, including their safety in the laboratory. It is rather absurd that Ivor's consciousness on this issue was directed towards his own self-interest rather than the students' interests, although the two might not necessarily always be in conflict with each other. The above excerpt adds weight to the assertion that there are close relationships amongst Ivor's low level of professional consciousness, low level of professional confidence and his 'readerly' interpretation of the TAS regulations.

To conclude, for Ivor, despite his stated desire to improve his teaching and his students' learning of the subject matter, the introduction of the TAS with its many regulations was seen by him as imposing severe constraints upon his professional autonomy. Under such circumstances, teacher professionalism was severely compromised as Ivor struggled to make sense of his changing roles and responsibilities both as an assessor and a teacher. Most importantly, underpinning his 'readerly' interpretation of the TAS regulations was his consciousness of "protecting himself" rather than defending his students' interest.

2. CARL : THE BENEFIT OF DOUBT SHOULD GO TO THE SUSPECT

Carl was one of the teachers who exhibited the highest level of professional confidence and ability to maintain a proactive role in terms of managing the changing demands made upon him by TAS. With this high level of professional confidence, Carl was able to impose his own professional interpretations on the TAS regulations, to balance its demands against other professional priorities, and to exploit to the maximum what remained of his professional autonomy. For example, Carl allowed his students to look up references during the practical, a practice that was seen by many teachers as violating the TAS regulations. When asked if this would pose a dilemma for him in coping with the TAS regulations, he responded in the following way.

> No, it poses no dilemma to me at all. Because there is not any statement in the TAS Handbook which states that references are not allowed in the practical. Looking up references is what scientists do day in and day out. I just worry that students do not look up references. It is so good if they want to do so. I think this is an important skill that students should acquire. I think this has to be emphasized in fact.

Similarly, when asked to comment on the extent to which he allowed his students to confer freely amongst themselves during the practical (which other teachers would normally avoid), Carl once again interpreted the TAS regulations in a 'writerly' manner.

> I am aware of the conflict between assessment and teaching… But there is no formal statement about the DO's and DON'Ts in the TAS handbook…

Carl's attempt to create the greatest amount of room for himself to manoeuvre within the framework of the TAS and to be 'in control' of his own teaching is further revealed in the following interview excerpt.

> The TAS Handbook does not prohibit teachers from responding to students' questions during the practical. I have never found a statement there which prohibits a teacher from responding to academic questions raised by students. *I think, according to the rule of law, if it is not stated clearly, the benefit of doubt should go to the suspect.*

Undoubtedly, as implied from his use of the metaphor of "the rule of law", Carl was very confident about his own interpretation of the TAS regulations. Underpinning Carl's professional confidence was his 'scaffolding' view of learning, as revealed in his readiness to initiate discussion with individual students on the experimental results they obtained during assessment practicals, a practice other teachers normally avoided.

> I think it is important to do so. First, I just initiate the discussion. Initiate! Second, this is to focus the students' attention on the important points … It is just a problem of guidance, a sort of moral support, nothing more than that. It is quite a good gesture in this kind of situation and is also natural enough.

Besides being conscious of facilitating students' academic learning, Carl's professional consciousness was also directed towards providing students with an all-round education. This was discussed in the previous chapter in great detail. To recap briefly, Carl was problematising the situation of extending the assessment practical beyond normal school hours in order for teachers to be able to supervise their students' work, as required by the TAS. He called into question the educational value of such a practice, in particular, with regard to the provision of an all-round education for students.

In sum, Carl was able to distance himself from the details of his classroom practice by calling into question the taken for granted assumptions and values, and to problematise his own practice by asking questions such as: Who benefits? Who suffers? Whose interests are being served? Professional consciousness of this kind definitely contributed to Carl's high level of professional confidence and, in turn, his professional interpretation of the TAS regulations. Put simply, the most significant influence on Carl's teaching was not the formal apparatus of external obligation and control imposed on him by the TAS regulations, but his personal sense of professional obligation to offer students an all-round education. In other words, Carl exhibited "extended professionalism" (Hoyle, 1974), which takes account of a broader educational context and a wider range of professional activities.

3. DAWN : I HAVE EVOLVED… I AM MUCH MORE LIBERATED NOW

Dawn was another teacher who exhibited a high level of professional confidence and, hence, a professional interpretation of the TAS regulations; as she put it:

> I interpret the TAS regulations in a way which is quite different from other teachers.

Dawn's 'writerly' responses to the TAS regulations is reflected in her response to a situation where she allowed her students to discuss in groups while they were drawing up their experimental proposals – a practice that she did not feel was violating the TAS regulations, as some teachers might.

> It doesn't matter because this is a type II assessment, which allows the teacher to give guidance to students. A very wide range of guidance could be given to students. It is just a matter of me – the teacher – providing the guidance, or the students themselves providing the guidance to each other... In the past, I did not interpret the regulations in this way. That's why I said that I have evolved.

Indeed, as Dawn expressed it here, her 'writerly' responses to the TAS regulations was not evident at the outset of its introduction; it resulted from the experiences she had gained over the years.

> In the past, I felt very much constrained by the TAS. Now, I feel that I am set free again. This is an evolution really. I have evolved. At the beginning, I was very frightened... Now, I understand more about it, and I realize that I don't have to make it a bitter experience for me and for my students. In the first year of the TAS, I just did not know the flexibility provided by the TAS. I thought that it was something like an examination. I adopted a very strict attitude on every single item... The support provided by the TAS coordinator was very minimal. I did not have confidence in asking him questions. He often referred me to the Handbook. You know, where everything is set out in black and white. In that case, I had to work very carefully because this is an examination. In those days, most teachers had a very bad feeling towards TAS. I am much more liberated now. I like this word "liberated". I don't feel so constrained now. I don't mind students cooperating with each other as long as everyone is really working.

Thus, it is understandable why Dawn was not confident enough to impose a professional interpretation of the TAS regulations at its inception. When she actually expected collegial responses and suggestions, the TAS coordinator instead often referred her to the information set out in black and white in the Handbook. This lends support to the argument that where there are indications of heavy bureaucratic control, scope for professional development is generally reduced (Helsby and Knight, 1997). Indeed, Dawn attributed her evolution into taking on "a more liberated view" of interpreting the TAS regulations in subsequent years to her becoming a TAS coordinator herself. She said that conversations with like-minded colleagues provided opportunities for her to find out what practising the reform ideas might involve, and afforded her an opportunity to gain the insights of others on the practical problems of putting the ideas into actual practice. This lends support to the important role played by smaller collegial groupings in teachers' professional development and teachers' enactment of the reform by creating powerful incentives for teachers to revise their practice (Stigler and Hiebert, 1999). Peer influence motivated Dawn to reform her practice. Notwithstanding this, Dawn's 'writerly' responses to the TAS regulations were underpinned by educational beliefs. In particular, she emphasized the affective side of students' learning, as revealed in the following interview excerpt.

> There will be a bright teacher among any three cobblers; therefore, I always allow students to have discussions among themselves during the practical. I like them to behave in this way. I think they would feel very happy if they can come up with some sort of solutions among their peers. It is no good if it is always the teacher who tells them the solution... I think that this is a better way of learning... I believe that the class atmosphere has to be built up slowly. I mean the atmosphere where they help each other and love each other, sharing with each other... And this does not violate the TAS regulations.

Another frame of reference underpinning Dawn's 'writerly' interpretation of the TAS regulations was that "the TAS is for teaching, for learning" and that "learning is more important than assessment", as revealed in the following interview excerpts.

> I allow a certain degree of helping out among the students themselves. This would give them more confidence. I think the TAS is for teaching, for learning. As long as students do not actually help out each other by physically doing something for each other. That should be allowed.

> I don't mind students discussing with each other during the practical. It poses no dilemma for me at all. The ultimate thing is that if this is really meant to be a learning process, am I going to ask them not to say a single word? I just do it this way. I haven't look deeply into the regulations to see if this is allowed or not. I just think that students' learning is more important. For the same reason, I won't avoid giving assistance to students. I often take the role that I am here to give assistance. I am a teacher. So, I won't avoid it.

When the goals of facilitating students' learning and undertaking assessment were in conflict with each other, Dawn was able to find ways to resolve the tension in a proactive manner, as revealed in her response to the following situation when she asked her students to present to their classmates their proposed experimental designs. Altogether, four different methods emerged from the class discussion. When asked if there was a possibility that some of the students might change their mind after listening to others' presentations, thus affecting fairness of the assessment, Dawn responded in the following way:

> I just don't care too much about this. Firstly, I treasure very much their expressions of different sorts of ideas. Secondly, all of them listen to the same presentations. At the end of the day, each of them will have to choose for themselves which method they would like to work on. Anyway, I would take these presentations into account when marking their reports. I would mark their reports with a more demanding standard. And this will apply to all students.

Overall, with her consciousness directed to facilitating students' learning of the subject matter and building up a harmonious and collaborative learning atmosphere for them, together with the confidence gained through collegial exchanges with other TAS coordinators, Dawn was able to impose a professional interpretation of the TAS regulations. Thus, she was adept at finding spaces in which she could manipulate the TAS requirements with her own professional judgment. However, compared with Carl, Dawn could be said to exhibit "restricted professionalism", taking account of factors that are largely classroom-based and concerned mainly with learning of the subject matter.

4. ALAN : I AM TAKING ADVANTAGE OF THE TAS

Alan was the third teacher who exhibited a high level of professional confidence and, like the other two, demonstrated his professional interpretation of the TAS regulations. He was an "active user" of the reform (Firestone, 1989) who was able to make TAS work to his students' and his own interests. As he put it:

> I am taking advantage of the TAS. I use it as a motivation for students to do what I require them to do. I hope I can cultivate in them an intellectual hobby – a habit of thinking.

Unlike Ivor, who exhibited a low level of professional confidence, Alan did not submit easily to the viewpoint of the TAS coordinator:

> The TAS coordinator was not happy with the way in which I had carried out this practical with my students. He said that I didn't tell students the exact method to use, and hence had made the data very confusing and that some students could not get any result at all. But I told him that not getting the result was the result itself.

Backing up Alan's professional confidence was his understanding of the relevant educational theories behind his statement of "not getting the result was the result itself". In this case, the philosophy of science about which Alan was very knowledgeable indeed (see Chapter 4). Indeed, what distinguishes active users may not be solely that they have what it takes to do what reformers ask of them, but they have a threshold level of knowledge and skill that enables them to access the resources needed to continue to learn about new practice, especially when those resources are not readily available (Spillane, 1999). Below is another example demonstrating how Alan's professional confidence was backed up by his expert knowledge. In his response, he was referring to a situation where a student asked him if the graph for the experimental results should be drawn in the form of a straight line or a curve, whereupon Alan encouraged the student to consult his classmates' data before making a decision for himself.

> If a student can ask a sensible question, why do we have to block his development? I think his question is a sensible one – should it be a straight line or a curve? One could arrive at different conclusions depending on the type of graph drawn. He was not sure whether his set of data was accurate or not. He could be surer about it if he could make a cross-reference to others' results. TAS allows students to pool data. The TAS allows students to use other people's data if they could not get their own set of data. Right? (Indeed, Alan was right on the above two points!)

When interpreting the TAS regulations, Alan's consciousness was often directed towards adjusting his pedagogy to the nature and abilities of his students. For example, he didn't mind discussing with students the results they had collected from their experiment before they wrote up their reports.

> I discovered that for average performers like my students, even if you have initiated the discussion with them, they may not necessarily be able to develop it further. Even if they can develop it further, they may not necessarily be able to present their discussion on paper in a systematic manner. That is, there is a

> long way to go from initiation of discussion to their production of a decent report. This has to be accomplished by the students themselves. What I can do is to turn their heads around and ask them to look. Look. There is something here. You'd better be aware. You'd better be thinking along this line.

Thus for the same reason, Alan did not avoid discussion with students or giving assistance to them during the practical. He also disliked the idea of deducting marks from students for offering help to them, as some teachers normally did, because his professional consciousness was directed towards maintaining his students' interest in their study.

> Given the nature of my students, I don't want to 'kill' the students. Why do I have to be so cruel as to insist on deducting marks from them. This is to 'sentence the student to death' once again! This is a de-motivation to their learning. This would just remove the last bit of motivation there is with them.

This was in line with his view of himself "not as an examiner representing HKEA" but a teacher standing on the side of the students helping them "to cross over the hurdle of the TAS". Overall, Alan exhibited a high level of professional confidence. His 'writerly' responses to the TAS regulations had been guided by his consciousness towards adjusting his pedagogy in relation to the nature of his students, in particular, in maintaining their interest in their study.

5. EDDY : I CONFESS THAT I AM NOT HONEST

Many of Eddy's classroom actions were very much the same as those of Carl. For example, both of them allowed students to work in groups and to finish writing up their reports at home, which other teachers would normally avoid. In addition, Eddy distributed the lab manual to students in advance and conferred "a little" with them before coming to the practical.

How were Eddy's classroom actions related to his interpretation of the TAS regulations? Did Eddy interpret them in a 'writerly' manner as Carl did? At times, Eddy did put forward sound educational reasons to back up his actions and thus gave the impression that he was also interpreting the TAS regulations in a 'writerly' manner as Carl was. For example, on the issue of allowing students to work in groups, Eddy responded in the following manner:

> It is impossible to carry out scientific studies on your own. You have to cooperate with others. I often emphasize to them that they must have a division of labour for group work. I think this is very important. I was still able to differentiate the differences in their abilities though they were using the same set of data. Some of them presented in a better way than the others. It is perfectly acceptable to me to ask students to work in pairs.

However, a thorough analysis of all the data collected at different times of the study calls into question the interpretation that Eddy was imposing a professional interpretation on the TAS regulations. In particular, Eddy expressed on several occasions a sense of powerlessness and resignation as a result of the demands made on him by the TAS, which can be regarded as a typical 'readerly' response to policy texts.

> TAS sometimes makes me feel that both my students and me are sort of doing some homework. We have to carry out certain experiments purely for the sake of satisfying the TAS requirement.

Most importantly, Eddy's sense of powerlessness and resignation as a result of the demands made on him by TAS also manifested itself in his admission, in the following three instances, that he might have indulged in unprofessional practices. Firstly, on the issue of distributing lab manuals to students in advance of the practical work:

> I usually give students the lab manual one week in advance. Sometimes, I will tell them well in advance. We also talk a little about the practical first. Maybe this is somewhat unfair [to students in other schools] if it is for assessment…

Secondly, on the issue of allowing his students to finish writing up their reports in the library without his supervision, even though he was available to supervise:

> I confess that I am not honest when I ask students to finish their reports in the library. Once they get out of my sight, I don't know what they are doing. In fact, I am cheating somewhat. I pretend not to be able to monitor them because of my other engagements.

Thirdly, on the issue of allowing students to complete the report at home:

> Though I tell them to put their reports on my desk before they leave, in fact, they hand in their reports on Monday morning. I just pretend that they have handed them in to me already before they leave the school on Friday. That seems deviating a little from the TAS regulations.

One might simply infer from the above interview excerpts that Eddy exhibited a very low level of professionalism in allowing himself to indulge in such unprofessional practices. However, an important question emerges when one tries to compare Eddy's case with that of Carl. Though many of their practices were very similar, for example, both of them allowed students to work in groups and to complete their reports at home, their interpretations of the relevant TAS regulations differed markedly. Whilst Eddy thought that he might be "cheating", Carl didn't have the slightest sense of indulging in any unprofessional practices, as reflected in statements like "There are no DO's and DON'Ts in the TAS Handbook". In short, Eddy interpreted the TAS regulations in a 'readerly' manner and Carl in a 'writerly' manner. Why was it so? What was the main contributory factor in the different interpretations of the same regulations?

As pointed out earlier, professional interpretation is associated with a high level of professional confidence, which is, in turn, guided by the teacher's professional consciousness. Was this the case for Eddy? The following interview excerpt provides some indication of the answer.

> I am very concerned to gain students' confidence in me. Thus, I give them a year plan at the beginning of the term. This will give them an impression that I am well prepared for their lessons. They will pay more attention to your lessons if they think that you are well prepared.

Taken by itself, "gaining students' confidence in me" appears to be quite a reasonable explanation for the use of a year plan, aside from its normal function in long term planning of the expected learning outcomes. Additionally, from the above quotation, one would expect Eddy to have spent much time and effort in planning his lessons. However, that seemed not to be the case because three out the four practicals observed had turned out not to be 'working well'. He confessed that he had not done any pre-lab preparation at all. In a later interview, when asked on what his considerations were based when drawing up the year plan, Eddy admitted that he did not actually look into details of what the expected learning outcomes would be. Rather, he was just sort of "rushing to finish the year plan to fulfill the administrative requirement of the school authorities". So, it could be argued that Eddy's consciousness was directed towards 'satisfying the demands of the various stakeholders at a superficial level', such that he would be able to maintain a good image of himself as a teacher, both from the perspective of his students and from that of the school authorities, instead of honestly addressing the goal of facilitating students' learning as a professional teacher ought to do. The following interview excerpt adds weight to this assertion. Eddy was responding to the question of why he continued to ask every student to carry out an individual project, even though the TAS had done away with that requirement:

> I fear that the principal may ask me why I don't continue this practice. In that case, I don't know what to say. I told her in the first year that this was a TAS requirement when I invited her to the project presentation session. After the presentation, she praised me a lot, saying that the effect was very good. She reported this during the staff meeting and invited other colleagues to consider the idea. A year later, my partner followed suit when the TAS did not have such a requirement any more. There has been some sort of peer pressure since then. Therefore, it seems not so good if I discontinue the practice.

So, the prime concern for Eddy in continuing the practice was not for its beneficial effects on students' learning, but rather to enhance his own image before the principal and his colleagues even though the two goals might not necessarily be in conflict with each other. In sum, evidence suggests that Eddy's consciousness was directed to his own self-interest rather than his students' learning. This, in turn, led to his low level of professional confidence and hence, to his seemingly 'writerly' but actually 'readerly' interpretation of the TAS regulations.

Why do I impute a 'readerly' interpretation on Eddy's practice? In essence, because Eddy didn't feel in his heart (i.e. his consciousness) that he was doing something that was professionally defensible and sound. Instead, he felt that he was indulging himself in some unfair practices and confessed that he was cheating. Thus, in a sense, Eddy was interpreting the TAS regulations in a way that was very similar to Ivor's. The only difference was that Ivor "followed the regulations closely" and hence did not have the sense of guilt, but only a sense of "fear" and "worry" about breaching the regulations. On the other hand, Eddy interpreted the regulations in the same 'readerly' manner but acted contrary to them, resulting in the feeling of having cheated.

The very same reason explains why Carl, whose practices were very similar to Eddy's, did not have the slightest sense of "cheating", but insisted that "the benefit of doubt should go to the suspect". That is, Carl had a high level of professional consciousness (in attempting to provide students with an all-round education) and professional confidence backing up his professional interpretation of the TAS regulations.

6. WHAT DO THE FIVE CASES TELL US?

Collectively, these five case extracts show that the state of being 'readerly' or 'writerly' was not inherent within the TAS regulations. Rather, it was dependent upon the teacher concerned. In other words, it was a matter of the teacher's personal choice, guided by his/her own beliefs and personal theories about teaching and learning. Table 14-1 summarizes the professional situations of the five teachers in regard to their interpretation of the TAS regulations, their professional confidence and professional consciousness. The teachers are arranged in descending order according to the professional confidence they exhibited. The interview excerpts included in the table have all appeared in one form or another in this or earlier chapters.

From the table, it is clear that those teachers who were more concerned with their own self-interests, like Eddy and Ivor, would have low confidence and tended to have 'readerly' responses to the TAS regulations. Consequently they felt that the TAS imposed many constraints on their teaching. In contrast, teachers like Dawn, Alan and Carl saw the TAS as an opportunity to reconstruct their pedagogy, such that their students' learning and welfare could be enhanced in one way or another. This was effected through their 'writerly' interpretation of the TAS regulations, backed up by their professional confidence which was, in turn, guided by their consciousness of enhancing students' learning (e.g. Alan, Carl and Dawn), and providing students with an all-round education (e.g. Carl), or learning other life skills (e.g. Alan).

Of the five teachers, Alan and Carl were the two who were able to distance themselves from their day-to-day practice and the teaching of the subject matter itself. They were able to direct their consciousness towards values outside the remit of the TAS, for example, the provision of an all-round education in the case of Carl and the cultivation of an intellectual hobby – a habit of thinking – in the case of Alan. Their strong commitment to their respective goals (i.e. their professional consciousness) fueled them with the professional confidence (manifested in Carl's use of the metaphor of 'the rule of law' and Alan's insistence of "getting no result is the result itself") to interpret the TAS regulations in a 'writerly' manner. Table 14-2 summarizes the professional situations of the remaining five teachers in this study in terms of their interpretation of the TAS regulations, their professional confidence and professional consciousness.

Table 14-1. Teacher's professional confidence and their interpretation of TAS regulations (I)

Teacher	Interpretation of TAS regulations	Professional confidence	Professional consciousness
Alan	***'writerly'*** responses	***High*** – I am *taking advantage* of the TAS… I told the TAS coordinator that not getting the result was the result itself.	***Students' interests*** – I see my role as a students' companion. I am with them *to cross over the hurdle of the TAS*… I want to *cultivate in them an intellectual hobby, a habit of thinking*… This is to *let them behave like scientists*… I want them to *learn to cooperate* among themselves... These are some techniques that are required in *real life*…
Carl	***'writerly'*** responses	***High*** – There is no formal statement about DO's and DON'Ts in the TAS Handbook… The *benefit of doubt should go to the suspect*… It poses no dilemma to me at all.	***Students' interests*** – The TAS aims to integrate assessment with teaching and *learning*… Is it really fair if students are denied opportunities to participate in *extracurricular activities?*… This is impossible when we are emphasizing *all-round education.*
Dawn	***'writerly'*** responses	***High*** – I have *evolved.* I am much more *liberated* now… I don't want those terrible things in the first year of the TAS to happen again…	***Students' interests*** – I want them to *learn* and to *share*… Biology is *fun*… If I don't do this, how are they going to *learn*? … Teaching without learning is nothing… The TAS is for teaching, for *learning*…
Eddy	Seemingly 'writerly' but actually ***'readerly'*** responses	***Low*** – I *confess* that I am not honest… In fact, I am *cheating* somewhat… I *fear* that the principal may ask me why I don't continue this practice …	***Teacher's self-interests*** – I am very concerned to gain students' *confidence in me*… In a way, I am quite *selfish.*
Ivor	***'readerly'*** responses	***Low*** – We have to *follow* the TAS regulations *closely.* I have to be *very careful* not to cross the line… I really have *no confidence* because it is easy to cross over the line.	***Teacher's self-interests*** – I have to *protect myself*… I would be in great trouble if there were accidents in my practicals… I have to behave like a policeman … *in order not to be scolded by the superior.*

Table 14-2. Teacher's professional confidence and their interpretation of TAS regulations (II)

Teacher	Interpretation of TAS regulations	Professional confidence	Professional consciousness
Bob	***'writerly'*** responses	***High*** – There is *no need* to avoid discussion with students because I am just sort of listening and approving their plans… There is *no need* to avoid discussion with individual students unless it comes to the discussion and interpretation of results.	***Students' interests*** – They are *learning* while I am assessing them… I'm a teacher with a mission… who cares about his students in *all aspects of their development*… I replace one of the regular tests with the individual project which, I think, contributes significantly to students' *learning*.
Fay	***'readerly'*** responses	***Low*** – If I were to teach some even brighter students … I *fear* that I may not be able to inspire them… If there is no TAS, I would be more at ease (to discuss with the students)…	***Student's interests*** – It is not fair if they still look up references at this stage… I decided not to tell them that they had set up the experiment wrongly … otherwise this would make the *assessment* meaningless…
Glen	***'readerly'*** responses	***Low*** – I *fear* that they would misunderstand *me* that I don't care about them.	***Students' interests*** – I care about students' *academic performance* and I put pressure on them… I just wanted to remind her that she shouldn't lose a larger portion (the *marks* for the report) for the sake of trying to gain a little portion (the *marks* for the experimental skills). In a way, I'm protecting *her* interests… I prefer Ss making use of the summer vacation to revise their lessons to that of carrying out an individual project… It'd be a waste of time.
Hugo	***'readerly'*** responses	***Low*** – *I don't think* as an individual *I can do much* to improve the situation unless … the school authority can think of some practical methods to change their [the students'] way of learning.	***Students' interests*** – Learning in school is similar to working in society… I was telling students the facts of life… They have to *learn to adjust to real life situation*… When compared with knowledge in books. These … are more practical and *useful to students*.
John	***'readerly'*** responses	***Low*** – I *fear* that she will go home and complain to her mother … *I* would then be *in trouble*… *I* have to be *very cautious* … to give them an impression that I am very fair in the process…	***Teachers' interests*** – *My* work is always driven by examinations… There are several things that one would look for when one is assessing a *teacher's performance*… his students' external examination results … whether he is a marker or a member of the subject committees in the public examining body.

6.1 Professional consciousness and professional confidence

At the heart of this chapter is an attempt to understand how the ten biology teachers explained their TAS practices in relation to their interpretation of the TAS regulations – the base upon which their professionalism is inferred. Whilst all teachers in this study were constrained to a greater or lesser extent by the regulations stipulated in the TAS Handbook, some teachers, like Ivor and Fay, appeared ready to follow an external prescription, even when they judged it to be misguided. There was a sense of powerlessness and resignation. At the other extreme, there were some examples of continuing professional confidence, where teachers like Carl, Alan and Dawn treated the TAS regulations as 'writerly' texts and sought to integrate its requirements within their own professional autonomy, compelling the TAS to work in the interests of their students. This suggests that teachers are able to exercise control of their own teaching by adopting a critical stance to policy change. The variations in how teachers adopt such a critical stance are a function of their professional confidence, which, in turn, is guided by their professional consciousness. What then is the nature of professional consciousness?

There appeared to be a complex of factors at work, including those personal and situational factors suggested by Woods et al. (1997), who pointed out that not all teachers are equally committed to teaching, nor in the same way. Similarly, Sikes et al. (1985) identified four different forms of commitment to teaching among teachers, namely, vocational, professional, instrumental and political forms of commitment. They claimed that teachers might have some or all of these to varying degrees. Data from this study, however, show that teachers can be further delimited into two groups, according to whether they are conscious of protecting students' interests or protecting their own self-interests in the very moment of their teaching. Given that teachers are often regarded as professionals who need "to act on judgments made in the best interests of the clients" (Hoyle and John, 1995, p.77), such an empirical finding of teachers' consciousness during their teaching could therefore best be referred to as their professional consciousness. The data also show that a teacher's professional consciousness has a strong bearing on what stance he/she will adopt towards a policy change, and whether he/she will be able to exercise control of his/her own teaching, compelling it to work in the interests of the students.

The case extracts also suggest that whilst there was scope for teachers to adopt a 'writerly' approach to the prescriptive TAS regulations, some failed to assert their professionalism in this way because of a lack of confidence. These teachers, like Eddy and Ivor, tended to adopt a largely reactive, rather than proactive role in their responses. This suggests a need for greater teacher support and more professional development in this respect than is currently available. In particular, there is a need to raise teachers' awareness of the de-professionalising aspects of their work (Hargreaves, 1994). This presupposes skilled and confident teachers who need time and space to reflect and to question values, something that is difficult to achieve in light of the intensification of teachers' work and the increasing trend of short term task orientation of in-service training. Short courses that focus on survival strategies and 'tips for teachers' are unlikely to stimulate the quality of thinking and reflection necessary for change and development. For teachers to regain their professional confidence and play a significant role in curriculum reform, several key implications for teacher education have to be considered. In particular, the important effect on professional learning, reported by Dawn, of sharing experiences with like-minded colleagues suggests a need to develop a very broad notion of professional development. The concepts of lifelong learning and continuous improvement (as well as professional updating) point to the need to embrace a variety of forms of learning, of which initial and in-service teacher education constitutes only one part (see Chapter 15). What will be of crucial importance is the extent to which demoralised teachers, like Ivor, can regain their professional confidence, and reassert their professionalism in terms both of their critical reading of central policy texts and of their assumption of a more proactive role in educational reform. As Hargreaves (1994) argues, teaching in the present age is characterized by struggle. There are forces and groups intent on deprofessionalizing teachers and their work, whereas there are others seeking to redefine teachers in more positive ways. Teachers themselves occupy the key position in this conflict. It is important, therefore, in the pursuit of enhanced professionalism, that teachers engage with changes, rather than be taken over by them. In order to do that, they need to understand the origins and nature of the changes, and their own responses to them. This study is a contribution towards that endeavour.

Chapter 15

WAYS OF SEEING AND WAYS OF ENACTING

In its broadest sense, the study reported in this book examined the relationship between the teachers' acts and their underlying beliefs in the specific context of the HKAL Biology Teacher Assessment Scheme (TAS). The study addressed the following research questions:

1. What are the characteristic features of secondary science teachers' classroom actions in the TAS?
2. In the context of classroom teaching, what personal understanding / perceptions do secondary science teachers have of the TAS?
3. What are secondary science teachers' beliefs about science, teaching and learning?
4. What are the relationships among teachers' beliefs about science, teaching and learning, their understanding of the TAS reform, and the ways they implement the reform inside their classroom?

Details of the answer to the first question were reported in Chapter 3 in the form of a combined account of the teachers' practices in the TAS. Answers to the second and third questions were reported in Chapters 4 to 11 in the form of individual case stories for the teachers concerned; and the cross case analysis reported in Chapter 12. The most salient points raised by those three questions will be used here to answer the fourth question. Related conclusions are also presented. In addition to the conclusions of the study, this chapter also addresses its implications for (1) teacher education and teacher professional development, (2) educational reform, and (3) future research.

As pointed out earlier, in order to answer the fourth research question, I need to make reference to some of the findings reported in earlier chapters. The following five sub-sections serve this purpose, and each contributes to providing an answer for the fourth research question.

B.H.W. Yung, (ed.), Assessment Reform in Science, 205-226.
© 2006 *Springer. Printed in the Netherlands.*

1. CHARACTERISTIC FEATURES OF TEACHERS' CLASSROOM ACTIONS IN THE TAS

Though there were striking differences among the teachers' practices, there were also comparable tendencies amongst them. In particular, the following comparable tendencies amongst the teachers are worthy of reiteration.

First, there were appreciable differences among the teachers in terms of the relative proportion of T-S interaction they devoted to assessment-related issues. This ranged from the lowest of less than 1% of the total DTUs in Alan's class, to the highest of more than 13% of the total DTUs in John's class (see Figure 3-3). The first row in Figure 15-1 summarizes the situation by arranging the ten teachers in order of the *increasing* proportion of T-S interaction spent on assessment-related issues.

Second, there were also appreciable differences among the teachers in terms of the amount of T-S interaction they had with their students during the TAS practicals. This ranged from the highest figure of 718 DTUs per practical in Alan's case, to the lowest of only 132 DTUs per practical in John's class (see Figure 3-2). The second row in Figure 15-1 summarizes the situation by arranging the ten teachers in order of their *decreasing* amount of T-S interaction with their students during the TAS practicals.

Third, notable differences were found among the teachers in terms of the proportion of T-S interaction spent on discussion of concepts underlying scientific research (see Figure 3-5). This ranged from the highest of 53% of the DTUs related to the teaching of practical work *per se* in Bob's class to the lowest of only 15% in John's class. The third row in Figure 15-1 summarizes the situation by arranging the ten teachers in order of the *decreasing* proportion of T-S interaction spent on discussion of concepts underlying scientific research.

Fourth, besides the amount and foci of T-S interaction, there were also notable differences among the teachers' *patterns* of interaction with their students. In particular, the teachers varied a lot in terms of the proportion of T-S interaction used to guide the student(s) to an answer rather than telling them the answer directly (see Figure 3-6). This ranged from the highest of 83% of the DTUs related to teaching of practical work *per se* in Bob's class to the lowest of only 19% in John's class. The fourth row in Figure 15-1 summarizes the situation by arranging the ten teachers in order of the *decreasing* proportion of T-S interaction used to guide students to the answers rather than telling them the answers directly.

Increasing proportion of T-S interaction on assessment-related issues →

Alan	Bob	Carl	Dawn	Eddy	Fay	Glen	Hugo	Ivor	John

Decreasing amount of T-S interaction per TAS practical →

Alan	Bob	Carl	Dawn	Eddy	Glen	Ivor	Hugo	Fay	John

Decreasing proportion of T-S interaction on concepts behind scientific research →

Bob	Alan	Carl	Dawn	Eddy	Ivor	Fay	Glen	Hugo	John

Decreasing proportion of T-S interaction with T guiding S to answers →

Bob	Alan	Dawn	Carl	Eddy	Fay	Hugo	Ivor	Glen	John

Figure 15-1. Comparable tendencies among the teachers in their implementation of the TAS

2. TEACHERS' UNDERSTANDING / PERCEPTIONS OF THE TAS

Evidence presented in the case stories points to the fact that the teachers in this study understood and perceived the TAS in rather different ways. One group of teachers saw the TAS primarily as an assessment reform aimed at differentiating students' practical competence very accurately. Another group of teachers saw the TAS as an assessment reform *with* a pedagogical dimension advocating the teaching of practical work using an investigative approach. Figure 15-2 summarizes the situation of the ten teachers regarding their understanding / perceptions of the TAS.

TAS is an assessment reform that also has a pedagogical dimension					TAS is an assessment reform aimed at accurately differentiating students				
Alan	Bob	Carl	Dawn	Glen	Eddy	Fay	Hugo	Ivor	John

Figure 15-2. Teachers' understanding / perceptions of the TAS

Teachers who perceived the TAS primarily as an assessment reform aimed at differentiating students often saw the purpose of the whole business as producing periodic, cumulative evidence for substantiation of the TAS marks to be submitted to the public examination authority. They saw the

TAS assessments as separate from, and even in opposition to, a more routine monitoring of progress, which also involved responding to students' learning problems as they were encountered. The thrust towards formality was largely taken for granted, and the reliability and comparability of the TAS assessments were a much greater concern than their validity. As such, the orientation was towards producing summative assessments for a third party, the public examination authority, rather than formative assessments to support learning.

These findings largely corroborate similar studies on teachers' classroom actions in school-based assessment schemes, in that:

- teachers are much bothered by having to assume the dual roles of assessors and teachers (e.g. Donnelly et al., 1994; Paechter, 1995; Torrance, 1995b; Harlen and James, 1997);
- teachers become reluctant to help students (e.g. Torrance, 1991; Jones et al., 1992; Broadfoot, 1995); and
- teaching and assessment remain largely polarized (Peterson, 1992).

Yet a significant finding of the present study is that there were teachers (e.g. Alan, Bob, Carl and Dawn) who also paid sufficient attention to the pedagogical dimension of the assessment reform. These teachers identified the issue of how and in what ways TAS assessments should impact on learning as a priority for them, in addition to being an incremental source of assessment data. That is, they put due emphasis on the quality of T-S interaction and their feedback to the students during the course of such interactions. This resulted in qualitatively very different learning environments in the classrooms of these teachers, compared with those who saw the TAS primarily as an assessment reform.

3. TEACHERS' DIFFERENT VIEWS OF FAIRNESS

Also underlying the teachers' enactment of the TAS was their concern with fairness. Indeed, much of the data of the present study supports Dunne's (1999) findings that "the teachers developed techniques by which they distanced themselves from their assessment decisions" and adopted what she called a "positioned neutrality". Dunne explained that "this distanced professional positioning displaced responsibility from the teacher and the interactive context, such that a student's relative success or failure became a marker of their own individual attributes" (p.124). A significant contribution of the present study to this literature is, however, the identification of three different views of fairness held by the teachers. That is, the notion of fairness being interpreted in the context of:

- extension of the public examination procedures
- providing chances for students to learn the subject matter
- providing students with an all-round education

Figure 15-3 summarizes the different views of fairness held by the respective teachers.

Receiving an all-round education	**Learning the subject matter**	**Extension of public exam procedures**
Carl	Alan Bob Dawn Eddy Glen	Fay Hugo Ivor John

Figure 15-3. Teachers' different views of fairness

It has to be reiterated that the current study was not designed with the aim of identifying teachers' different views of fairness; rather, this was an aspect that emerged from the data itself. It should also be noted that data of this sort would not have been made available for further analysis if the study had not set out to study the teachers' practice from the *teachers' own perspectives* and in the context of how they *experienced* it.

4. TEACHERS' BELIEFS ABOUT SCIENCE, TEACHING AND LEARNING

The case stories indicate that the teachers had different views about science, the role of practical work, teaching and learning. Figures 15-4 to 15-6 summarize the situation for the ten teachers in each of the respective areas.

Affective / cognitive / motor skills	**Little thought about it / other instrumental goals**
Alan Bob Carl Dawn Fay	Eddy Glen Hugo Ivor John

Figure 15-4. Teachers' beliefs about the role of practical work

With respect to their beliefs about the role of practical work in school science, there were teachers who saw practical work as contributing to developing the appropriate cognitive/motor skills and/or motivating students' interest in learning science. There were also teachers who gave little thought to the role practical work in school science education (e.g. Eddy and Glen), and those who made use of practical work to achieve some instrumental goals like helping students to get better examination scores (e.g. Ivor and John), or preparing them for their future working life (e.g. Hugo).

Dynamic / Theory-laden	**Static / Objective**	**Amorphous view**
Alan Bob Carl	Dawn Eddy Fay Glen Hugo	Ivor John

Figure 15-5. Teachers' beliefs about science

With respect to the beliefs they held about science, the ten teachers could be divided into three groups (see Chapter 12). The first group of teachers saw science as a dynamic and on-going business, and believed that observation is theory-laden (e.g. Alan, Bob and Carl). The second group of teachers viewed scientific knowledge as truth that would remain unchanged throughout time and/or that science is very objective (e.g. Dawn, Eddy, Fay, Glen and Hugo). The third group of teachers held an amorphous or unarticulated view of science, which could not be easily classified into either of the above categories.

Learning-focused	**Teaching-focused**
Alan Bob Carl Dawn Fay	Eddy Glen Hugo Ivor John

Figure 15-6. Teachers' beliefs about teaching and learning

With respect to the beliefs they held about teaching and learning, the teachers in this study could be divided into two groups. One group of teachers held beliefs that were more learning-focused (i.e. their concern was that students really learnt what they were taught) whereas the other teachers held beliefs that were more teaching-focused (i.e. they focused more on their own "teaching" and assumed that students would learn whatever they were "taught").

Overall, the data from this study concur with other studies showing that the way teachers teach largely depends on who they are, and on their own history, experiences, educational visions and aspirations (e.g. Nespor, 1987; Louden, 1991; Wallace and Louden, 2000); beliefs about the nature of science and the role of practical work (e.g. Brickhouse, 1989, 1990; Fischler, 1994; Hashweb, 1996); as well as their beliefs about teaching and learning (e.g. Hewson and Hewson, 1989; Hawthorne, 1992; Tobin and McRobbie, 1997; Bryan, 2003). The current study found, in addition, that the notion of teachers' professional consciousness was of crucial importance in guiding these teachers' work, especially in the context of educational reforms.

5. PROFESSIONAL CONSCIOUSNESS, PROFESSIONAL CONFIDENCE AND POLICY INTERPRETATION

As with the uncovering of the teachers' different views of fairness, the notion of teachers' professional consciousness was not an object of research at the outset of this study, but was an aspect that emerged from the data itself. As revealed in Chapter 14, teachers in this study can be classified into two groups with respect to what their professional consciousness was directed towards. Figure 15-7 summarizes the situation for the ten teachers.

Protecting students' interest	Protecting teacher's own self-interest
Alan Bob Carl Dawn Fay Glen Hugo	Eddy Ivor John

Figure 15-7. Teachers' professional consciousness

The teachers whose professional consciousness was directed towards protecting their own self-interest (e.g. Eddy, Ivor and John) usually exhibited low professional confidence and hence *tended* to have 'readerly' responses to the TAS regulations. Consequently, they felt that the TAS regulations were imposing many constraints on their teaching. In contrast, other teachers saw the TAS as an opportunity for them to re-construct their pedagogy so that their students' learning could be enhanced in some way. This was effected through their 'writerly' interpretation of the TAS regulations and backed up by their professional confidence. This was, in turn, guided by their consciousness to enhancing students' learning (e.g. Alan, Bob, Carl and Dawn) and providing students with an all-round education (e.g. Carl). That is, this latter group of teachers were conscious of a need of protecting their students' interests.

Overall, the data of the present study are consistent with the general thesis in the literature that teachers who adopt a critical stance to policy change are able to exercise control of their own teaching, and that the variations in how teachers adopt such a critical stance are a function of their professional confidence which, in turn, is guided by their professional consciousness (e.g. Helsby, 1995; Proudford, 1998). The current study extends this relationship by pointing to the importance of teachers having their professional consciousness directed towards *protecting students' interests* rather than *their own interests* if they are to adopt a critical stance towards policy change.

6. THE STRUCTURE OF TEACHERS' AWARENESS IN RELATION TO THEIR ENACTMENT OF THE TAS

So far, I have been recapturing the main findings generated from this study. Are there any inter-relationships amongst them? To answer this question, all these findings can be put together as shown in Figure 15-8. For ease of interpretation, the teachers are divided into three groups, according to the proportion of their T-S interactions concerned with assessment-related issues, as follows:

- less than 2 % (e.g. Alan, Bob, Carl and Dawn)
- 4 to 6 % (e.g. EDDY, FAY and GLEN)
- 10 to 14 % (e.g. *Hugo*, *Ivor* and *John*)

For the sake of presentational clarity, the names of the teachers in these three groups are printed in three different font styles, as they appear above.

As evident from Figure 15-8, there was a *high correspondence* between the teachers' enactment of the TAS and their:

- perceptions / understanding of the TAS
- conception of fairness
- beliefs about science, teaching and learning
- professional consciousness

Figure 15-8. The different layers of teachers' awareness in relation to their enactment of the TAS

Implementation of the TAS

Decreasing amount of T-S interaction per TAS practical →
Alan Bob Carl Dawn EDDY GLEN *Ivor* *Hugo* FAY *John*

Increasing proportion of T-S interaction on assessment-related issues →
Alan Bob Carl Dawn EDDY FAY GLEN *Hugo* *Ivor* *John*

Decreasing proportion of T-S interaction on concepts behind scientific research →
Bob Alan Carl Dawn EDDY *Ivor* FAY GLEN *Hugo* *John*

Decreasing proportion of T-S interaction with T guiding S to answers →
Bob Alan Dawn Carl EDDY FAY *Hugo* *Ivor* GLEN John

Perceptions / understanding of the TAS

An assessment reform with a pedagogical dimension
Alan Bob Carl Dawn GLEN

An assessment reform aimed at differentiating students accurately
EDDY FAY *Hugo* *Ivor* *John*

Conception of fairness

Receiving all-round education
Carl

Learning the subject matter
Alan Bob Dawn EDDY GLEN

Extension of public exam procedures
FAY *Hugo* *Ivor* *John*

Beliefs about role of PW

Affective / cognitive / motor skills
Alan Bob Carl Dawn FAY

Little thought about it or other instrumental goals
EDDY GLEN *Hugo* *Ivor* *John*

Beliefs about science

Dynamic / Theory Laden
Alan Bob Carl

Static / Objective
Dawn EDDY FAY GLEN *Hugo*

Amorphous view
Ivor *John*

Beliefs about teaching & learning

Learning-focused
Alan Bob Carl Dawn FAY

Teaching-focused
EDDY GLEN *Hugo* *Ivor* *John*

Professional consciousness

Protecting students' interest
Alan Bob Carl Dawn FAY GLEN *Hugo*

Protecting teacher's own self-interest
EDDY *Ivor* *John*

In other words, the teachers' classroom actions in the TAS were very much related to their perceptions/understanding of the TAS, their conception of fairness, their beliefs about science, teaching and learning, as well as their professional consciousness. Based on these findings and capitalizing on Marton and Booth's (1997) idea of the structure of awareness (refer to Appendix A for details), a pictorial representation can be drawn up to make sense of the variations in how the TAS was enacted and experienced by the teachers in this study (Figure 15-9).

Before moving on, it must be reiterated that the pictorial representation is intended to serve as some sort of a metaphor for communicative purposes so that readers can better grasp the nature of, as well as the inter-relationships amongst, the various findings of this study. It is not intended to depict the thinking processes or mental representations of what might be taking place inside the minds of the individual teachers. In brief, Marton and Booth believe that the "inner" (beliefs) and "outer" (acts) of a person are not two things, nor is one held to explain the other. That is, the world is not constructed by the teacher, nor is it imposed upon him/her; it is constituted as an internal relation between them.

According to Marton and Booth, certain structures of awareness are implied by certain ways of understanding. A person is simultaneously aware of particular aspects of a situation or a phenomenon; and his/her awareness of these aspects implies a tacit awareness of other aspects. This means that certain aspects become figural, in focus or focal, whereas other aspects recede to the background. Hence the different shades of colour of the different layers in Figure 15-9 represent the different layers of awareness of the teacher. The darker the colour, the less likely that this aspect would be drawn into the teacher's focal awareness at the very moment of teaching a particular content. The broken lines separating the different layers of awareness illustrate their non-static nature. That is, the structure of the teacher's awareness may change as a result of the teacher encountering some new experiences, or vice versa. Thus, using such a pictorial representation, the world as experienced by the teachers in the context of the TAS can be depicted as follows.

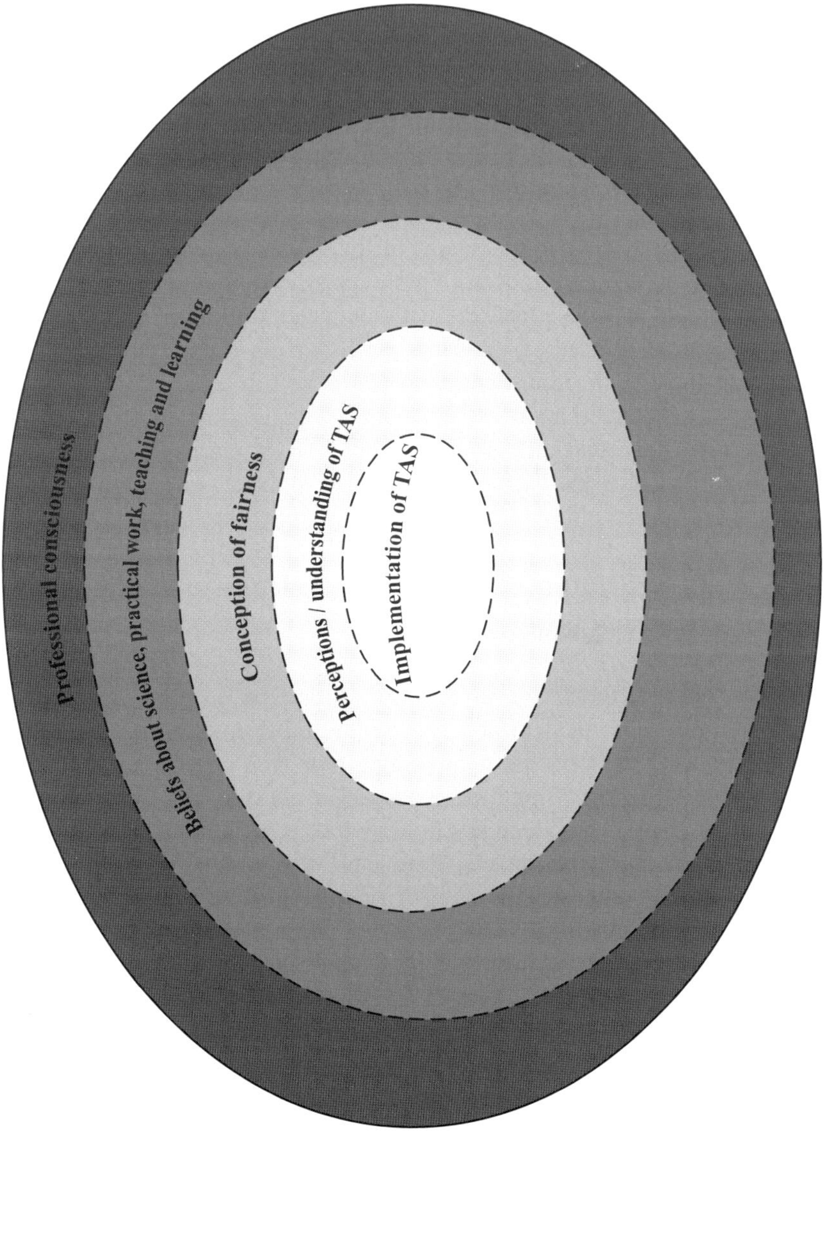

Figure 15-9. The structure of teachers' awareness in relation to their enactment of the TAS

6.1 Perceptions / understanding of the TAS

Embedded and *reflected* in the way in which a teacher enacted the TAS was his/her understanding/perceptions of the TAS. That is, whether the teacher saw the TAS primarily as an assessment reform (e.g. as in John's case) or one with a pedagogic dimension as well (e.g. as in Alan's case). These different perceptions/understanding of the TAS were reflected in the amount of T-S interaction and the foci and pattern of T-S interactions teachers had with their students (see Figure 15-8).

6.2 Conception of fairness

Underlying the teachers' enactment of the TAS and closely related to their perceptions/understanding of the TAS were also their conceptions of fairness (see Figure 15-9). Those teachers with a conception of fairness embedded in the context of providing students with an all-round education and/or providing them with opportunities to learn the subject matter tended to see the TAS as an assessment reform with a pedagogical dimension. On the other hand, those teachers who saw the TAS primarily as an assessment reform tended to view fairness in relation to what would be required in maintaining fairness in public examination settings (see Figure 15-8).

6.3 Beliefs about the role of practical work, science, teaching and learning

The teachers' different views of fairness can be seen as resulting from drawing into their focal awareness various other more *general* aspects of teaching during the very moment of their own teaching. These include their beliefs about the role of practical work, about science, teaching and learning (see Figure 15-9). Thus, teachers holding views of fairness in the context of providing students with an all-round education and/or providing students with the chance to learn the subject matter tended to have the following beliefs figuring in their focal awareness:

- *practical work* is for developing students' affective / cognitive / motor skills
- *science* is an on-going and dynamic business; observation is theory-laden
- teaching should be *learning-focused*; that is, concerned with whether *students* can really *learn* what they are taught

On the other hand, those teachers who saw fairness in terms of maintaining fairness of the assessment process tended to have the following beliefs figuring in their focal awareness (see Figure 15-8):

- *practical work* is for achieving certain instrumental goals like preparing students for the public examination or their future working life
- *science* is objective and scientific knowledge remains unchanged through time, etc.
- beliefs about teaching and learning that are more *teaching-focused*; that is, the teacher focuses on his/her own "teaching", assuming that students *will* learn whenever and whatever they are "taught".

However, which aspects figure in the teachers' focal awareness and which recede to the background were, in turn, related to the teachers' professional consciousness (see Figure 15-9).

6.4 Professional consciousness and policy interpretation

As pointed out above, crucial to a teacher's enactment of the TAS was also his/her professional consciousness; that is, whether the teacher was committed to protecting the *students' interests* or *his/her own self interests*. Teachers who were more concerned with their own self-interests (e.g. Eddy, Ivor and John) exhibited low professional confidence and tended to have 'readerly' responses to the TAS regulations. This was in contrast to those teachers who were more concerned with protecting students' interests, and saw the TAS as an opportunity to re-construct their pedagogy, such that their students' learning could be enhanced in one way or another. This was effected through their 'writerly' interpretation of the TAS regulations and backed up by their professional confidence, which was, in turn, guided by their consciousness of enhancing students' learning and/or providing students with an all-round education (e.g. Alan, Bob, Carl and Dawn). In other words, teachers' professional consciousness was very much related to what figured in their focal awareness and what remained in the background.

6.5 The structure of teachers' awareness

To sum up, there are various aspects of teaching, aims of education in general and science education in particular, beliefs about science, practical work, teaching and learning, the role of school-based assessment, the role of practical work and the particular content to teach, in relation to which teachers differ significantly. These aspects constitute different layers of the teachers' awareness, which are simultaneously present in teaching. In the practice of teaching (and assessing) a specific content, all these increasingly general aspects of the teachers' awareness are inherent. All these are present – admittedly not always clearly; not always explicitly – in every moment of the very teaching of the content. Of the various aspects of teaching, only those which the teacher sees as having relevance will be drawn into consideration in a specific instance of teaching; and these represent the directedness of the teacher's intentionality.

Though the pictorial representation in Figure 15-9 is constructed based on Marton and Booth's idea of the structure of teacher's awareness, the different layers of the teachers' awareness are empirically derived. As evident from the above discussion, analysing and interpreting teachers' experiences in the TAS using the pictorial representation as depicted in Figure 15-9 seems to make sense. Nevertheless, one outstanding question remains: How can teachers be helped in uncovering their own professional consciousness and in building up their professional confidence? This leads me to the implications of findings of this study for teacher education and professional development.

7. IMPLICATIONS FOR TEACHER PROFESSIONAL DEVELOPMENT

Analysing the data in the way detailed in this chapter has enabled me to gain a deeper understanding of the intertwining relationships among the teachers' beliefs, their aspired educational visions, their understanding / perceptions of the TAS, their classroom actions, their different views of fairness and their varying sense of professionalism. On the basis of this study, I strongly believe that initiatives to reform our schools will surely fail if they ignore the centrality of teachers' beliefs, in particular, those more central in their belief system. Teachers do not merely deliver the curriculum; they develop it, refine it and reinterpret it (Hargreaves, 1992). They are the ultimate key to educational change and school improvement, and educational change must always be mediated through the minds and motives of teachers, who are the final policy brokers (McLaughlin 1987, 1990). Systemic reforms are of little value if they do not take teachers into account. These conclusions point clearly towards the need for major investment in the continuing professional development of teachers.

This study also offers further evidence of the difficulty of promoting assessment as a formative, educationally based process in educational contexts, like Hong Kong, that stress measurement and accountability (Broadfoot, 1995). The high-stakes nature of the HKAL Examination may override the potential of assessment to foster meaningful learning (see John's case in Chapter 11). Harland's (1988) analysis of the de-skilling and technicisation of teachers through mandated policies is also relevant here. Ivor's case, reported in Chapter 10, was a case in point. The extent to which some teachers may have internalized the view that assessment is a matter of technical procedure rather than professional judgment is fully reflected in

Ivor's words, "We have to follow the TAS regulations closely." Yet it remains debatable whether we should then treat the teachers as no more than 'technicians', who merely need instrumental support, or as 'professionals' who need a broader programme of professional development that will allow them to apply their knowledge through exercising their own judgment within the TAS framework.

While the priority is to ensure that teachers can grasp the requirements of the TAS, two questions remain: Should teacher professional development be undermined by this instrumental priority? Should emphasis be laid on controlling and managing teachers through mandated prescriptions and close monitoring, rather than on developing and empowering them? To answer these questions, we need to make a distinction between two approaches to quality in education identified by Helsby and Knight (1997). One approach sees quality as something to be achieved through a system of bureaucratic accountability, signs of which can be seen in the data of the current study, which highlights some of the teachers' fear of not being able to comply with the expectations of the TAS coordinator (e.g. Ivor's case). In contrast, the other approach focuses on continuous quality improvement rather than on quality control. It is centred on the idea of trust and empowerment, and is believed to be more powerful and more appropriate to educational settings. The empowerment that lies at the heart of this second idea implies that teachers are seen as professionals, exercising judgment and creativity, rather than as technicians, following directions. Moreover, this approach, by definition, implies continuous learning, lifelong professional development.

If the observations laid out above are solidly based, and I believe they are, they hold some serious implications for teacher education and teacher professional development in general, and in the area of preparing teachers for more school-based assessments in particular. For instance, the three cases (John, Bob and Carl) reported in Chapter 13 are good illustrations of how different notions of fairness held by the teachers can result in approaches to science teaching that are qualitatively very different. They drive home the need for teacher professional development with respect to the role of teachers in school-based assessment, especially in terms of developing teachers' analytic awareness of the issue of fairness. Similarly, the five cases reported in Chapter 14 collectively point to the need to alert teachers to, or make them aware of, the professional consciousness behind their actions. Neither of these are easy tasks. The points which follow are developed from insights derived from data of the present study, and how they address the nature of teacher professional development and what this might entail.

Also echoed by data from this study is Sanger's (1990) notion that teachers' belief systems form deep layers of "calcified experience" (p.175). Changes rarely take place without destabilizing this deeper level of calcified experience. For this reason, challenges posed to teachers must be vigorous and explicit if change is to occur. For many teachers, it could be an awakening into consciousness, where suddenly the familiar daily routines of professional practice become discordant symbols of conflicts that existed between articulated and unarticulated levels of knowing. But the crucial question remains: What is it that enables such insight or sudden awakening to occur?

The case stories of teachers reported in this book contain concrete examples of their classroom actions, voices, concerns and methods used to solve practical problems, and information about the personal beliefs behind their classroom actions that were built up over a long period and established through the process of day-to-day teaching. A number of these teachers have provided examples of quality practical work teaching using an investigative approach in the context of a school-based assessment scheme. These will serve as good illustrations and models from which other teachers can learn. These examples, though in a sense idiosyncratic to the individual teachers with their own contextual variables, contain many teaching characteristics which are generic, just as there were some common threads in the beliefs and thinking of the teachers in these case stories that gave teachers' concern about fairness a dominant place in their teaching. Through seeing illustrations of these actual instances of teaching and carrying out school-based assessment, and of alternative ways of thinking, doing and assessing science from a *teacher's* perspective, other teachers may be able to extend their understanding and appreciation of the ideas and values of: (1) teaching science through the investigative approach; (2) assessing science through school-based assessment; and (3) how these had actually been put into action at the same time. Through reflection and evaluation of their own practices, teachers may be able to see beyond the specificity and idiosyncrasy of these classroom practices and use them to uncover their own professional consciousness. This may help them to re-organise their own belief systems and to re-direct their professional consciousness in relation to their own teaching contexts. This concurs with Putnam and Borko's (1997) suggestion that case teaching is particularly appropriate in preparing teachers for reform-based teaching because their opportunities to experience workable alternatives to conventional practice in actual classroom settings is likely to be quite limited.

But another question remains: How can the aim of transforming teachers' consciousness be sustained? The TAS has given rise to a number of in-service training courses which are predominantly instrumental. They are mainly one-off events where reform ideas are presented and teachers are offered no ongoing or sustained support for thinking about the reform ideas. Professional development is often framed narrowly as knowledge and skills-based, and improvement of teacher effectiveness is seen in terms of these specific training courses. Teachers are usually placed in a subordinate position in the planning and delivery of this kind of in-service training programme. Rarely is the agenda set by teachers, not even negotiated with them.

If the aim of transforming teachers' consciousness is to be sustained, the need for teachers to begin viewing themselves as true professionals, seeing themselves as contributing to the knowledge base that defines the teaching profession, and seeing that contribution as an integral part of what it means to be a teacher is prerequsite. Stigler and Hiebert (1999) claim that Japan has succeeded in developing a system of continuous school-based professional development that makes such goals achievable. One of the most common components of this school-based professional development system is "lesson study". In lesson study, groups of teachers meet regularly over long periods of time (ranging from several months to a year) to work on the design, implementation, testing and improvement of one or several "research lessons". Two features of the "lesson study", as reported by Stigler and Hiebert (1999), are relevant to the present discussion. First, the collaborative nature of "lesson study" provides a benchmarking process that teachers can use to gauge their own skills. *Collaboration* includes continuing interactions about effective teaching methods plus observations of one another's classrooms. These activities help raising the teachers' analytic awareness (see Chapter 13) by allowing teachers to reflect on their own practices in comparison with those of their colleagues. Second, "lesson study" not only helps teachers to develop but also develops knowledge about teaching that is relevant to classrooms and sharable among members of the teaching profession. In this sense, teachers participating in the lesson study begin seeing themselves as participating in development of the profession as a whole. They begin viewing themselves as true professionals.

To sum up, through the gradual improvement of individual lessons, and through the knowledge developed and shared during this process, the Japanese approach appears to have demonstrated its ability to ensure the steady and *sustained* improvement of teachers and teaching. This concurs with suggestions that teachers should be helped in developing *professional networks* in which they can discuss their work with one another, become teacher researchers, and write about their thinking. In Canada, Hodson and Bencze (1998) provided examples of the kinds of changes that such networks could bring about in the area of teaching of practical work. Models such as lesson study or the *teacher-as-researcher model* (Carr and Kemmis, 1986) will have to be implemented if we are to see change across the teacher population. This kind of professional development model enjoys the added advantage of giving teachers control over their own practice while generating changes. As Fullan and Hargreaves argue,

> Teacher development must actively listen to and sponsor the teacher's voice, establish opportunities for teachers to confront the assumptions and beliefs underlying their practices; avoid faddism and blanket implementation of favoured new instructional strategies; and create a community of teachers who discuss and develop purposes over time. (1992, p.5)

Research on teacher change (e.g. Richardson, 1990; Kagan, 1992) points to the inadequacy of older models of dissemination, and highlights several factors that inhibit adoption of innovations. Like other learners, teachers come to learn through social interaction with 'knowledgeable others' and/or their peers, through applying ideas in practice, and through reflection and the modification of ideas. Further assistance and support provided to teachers as they attempt to enact new practices in their classrooms is an essential and valuable component of efforts to help teachers to learn to teach in new ways (Black and Atkin, 1996). Support and feedback provided in conjunction with classroom observations of these attempts seem to be particularly helpful. These are found to be effective in enabling teachers to analyse, discuss, evaluate and change their own practices; in encouraging greater responsibility of their practices; and in fostering development of teachers' own theories of education (Calderhead and Gates, 1993). Thus, the next priority must be research on effective means of long-term, sustained and wide-spread teacher professional development in the area of school-based assessment. In this connection, New Zealand (e.g. Bell and Cowie, 2001; Gilmore, 2002) and the UK's experiences (e.g. Black et al., 2003) of providing professional development opportunities for teachers in the area of assessment are worth consulting. Creative uses of case stories like those reported in this book (and the metaphors contained therein) should also be actively considered (e.g. Yung, 2001).

8. IMPLICATIONS FOR EDUCATIONAL REFORMS

The findings of this study have one important implication for educational reforms, namely, the importance of getting teachers as well as the community at large involved in the process.

8.1 Getting teachers involved

As reflected in Alan's and Dawn's cases, teachers are motivated to proceed with their efforts to reconstruct their teaching when they find that students are interested in, and learning successfully with, the new teaching materials and new teaching approaches. Indeed, student success is a primary source of rewards for teachers (Lortie, 1975). However, rarely is this lever for change capitalized on by reformers (Spillane, 1999). Thus, reform processes which have not involved dialogue with teachers – such as the so-called 'teacher-proof' projects of the 1960s – have been less successful (Olson et al., 1999). If teachers' voices, such as those explicated in this study, could be heard in the process of change, this would remind reformers of the issues that are important to the teachers and of the intricacies which attend to proposed changes in classroom settings, and hence increase the chance of successful implementation of the reforms. Teachers' voices, as explicated in the current study, remind us that changes intended to enhance the quality of education may also be threatening and may pose considerable risks to the teachers concerned – words like "worry" and "fear" were often used in some of the teachers' descriptions of their work in this study.

On the other hand, teacher experience gained in the process of reform can itself lead to increased professionalism, allowing teachers to become better able to reflect on/analyze policy formation and the development of educational theory, as reflected in Alan, Bob, Carl and Dawn's cases. All such contributions are themselves an enhancement of the notion of teacher professionalism. Whilst the critical feedback which can come only from teachers is a vital ingredient of reform, the objective must be movement towards more open communication between teachers and reformers, based on greater collegiality in a common enterprise. On this basis, each party would benefit and the resulting synergy would strengthen the profession as a whole.

8.2 Getting the whole community involved

As implied in the above discussion, the success of any educational reform relies very much on the teachers. However, it will also be critically dependent on the context of the education reform in which the teachers have to operate. The enacted curriculum can be constrained by factors associated with the culture in which teaching and learning occur (e.g. Olson et al., 1999). Tobin and McRobbie (1996) documented how cultural myths held by science teachers constituted a referential system that favoured the retention of traditional practices. Of relevance to the current study are examples such as: time being a scarce commodity; a high priority goal being to cover prescribed content; a need to prepare students for the public examination and to do well in it. For instance, John's succinct remark – "My work is always driven by examinations" – is particularly illuminative of the situation. Similarly, it is understandable that Eddy perceived his principal's usual practice of reviewing students' performance in the public examination for each individual subject in the first staff meeting of the academic year as "a kind of invisible pressure". That is to say, John and Eddy's approaches to teaching and learning science were supported by cultural myths that had broad support among important stakeholders, including principals, parents and probably students. Accordingly, initiating reform involves more than working with individual teachers to change the ways they think and act in their classrooms.

This latter assertion is further supported by evidence from Alan's case. Alan accepted the rationale behind the TAS reform wholeheartedly. He invested a lot of effort in trying to make it work for his students. However, his effort was not appreciated by his principal who focused instead on a good "pass rate" for students taking biology in the public examination. In the end, because of his contrasting opinions with the principal regarding the goal of education, Alan resigned and joined another school. At the time of writing, I don't know whether Alan will find the culture of the new school more amenable to his own vision of education. Nevertheless, Alan's case is a good example of how a teacher's action may be affected by the sociocultural milieu in which he enacts his professional life. It also lends support to the gloomy scenario of the pervasiveness of cultural myths in supporting the status quo and constituting a conservative force in many of the recommended educational reforms. That is, teachers use myths, rather than the visions portrayed in the reform documents, as bases for deciding what makes sense and what does not (Tobin and Gallagher, 1987; Tobin et al., 1988; Tobin et al., 1994; Tobin and McRobbie, 1996). In such a situation, it is unlikely to be enough for teachers, like John, to change how they teach biology practical work, nor for critical masses of teachers, like Alan and Bob, to come together to spark chain reactions. Thus, it is imperative that approaches to reform are systemic and look beyond the efforts of individuals. As Black (1998) put it, in the context of assessment reforms,

> [R]eform has to be systemic in design, even if it may be piecemeal in implementation. Political strategies are bound to be affected by the understanding of assessment issues, on the part both of politicians and of the general public to whom they must appeal. (p.158)

Indeed, teaching and learning is embedded in a sociocultural matrix, and it is likely that education reform can only be initiated and sustained if the shared beliefs throughout a community support the recommended changes.

Students are important stakeholders in determining what happens in classrooms, and so too are a host of others including parents, colleague teachers, principals, university admission tutors and employers. If the reforms are to proceed as recommended, it will be necessary for discussions about teaching and learning to become a more visible component of the lives of all stakeholders, and for a majority to understand the need for change, advocate reform, reconceptualize their goals and associated roles, and support others in their efforts to enhance students' learning. It follows that reform of curriculum, of pedagogy, of assessment practices, all have to go together – reform of assessment cannot proceed on its own. Certification and accountability policies should be as consistent with, and as supportive of, such a reform as possible.

9. WHERE DO WE GO FROM HERE ?

Overall, the findings of this study suggest that good school-based assessment can be a powerful tool for improving teaching and learning (e.g. Alan's case). But in some cases, it is underdeveloped and its underlying rationale misunderstood by teachers (e.g. John's case). There is therefore a tremendous opportunity for improvement. If we are to realise the full potential benefits of school-based assessment, then more investment in research and development in this area are needed. Most important of all will be the research that provides greater insight into how teachers become reflective about their own assessment practices inside the classroom. That knowledge will help us learn what must be done in relation to the effective professional development of teachers. Until this is realised, technical solutions will continue to be seen as more conducive of change than teachers' professional learning and, in particular, teachers' reflection on their own practices.

In other words, teacher professional development must be understood as something far more extensive than participation in in-service or pre-service programmes. This implies that research into this area needs to investigate informal professional development as well as formal provision. In addition, such research needs to take into account the *experiential* aspects of teachers' work; that is, to explore professional development *as seen by individual teachers* with their own biographies, values, identities and educational aspirations. This research needs, therefore, to be *holistic*, located within a life histories approach, and to look at the process of professional development through *longitudinal studies* that attend to the teachers' *voices* (Goodson, 1997). The case studies reported in this book (in particular, Alan's and Dawn's cases) begin to capture something of the process, but they also make us aware of its complexity. As such, I believe that the major contribution of the current study to the literature is that it lays a solid foundation for further exploration into teacher professional development *in* and *through* assessment reforms. It has also demonstrated the fruitfulness of studying teachers' professional development in the context of assessment reforms, as recommended by McLaughlin (1997):

> Reforms of all stripes agree that teacher involvement in developing assessments and in assessing student work comprises perhaps the *single most potent opportunity of teachers' learning and change*. Engaging teachers in conversation about what students should know and be able to do, and how that performance could be assessed, prompts teachers to look critically at their own work and question the relationship between teaching practices and student learning. (my emphasis, p.83)

All in all, researching teachers' professional development in the context of assessment reforms becomes especially important given the growing international awareness of the need for changes in assessment practices (Broadfoot, 1995; Hamilton et al., 2002) and the range of significant initiatives that have come with it.

Appendix A

A STUDY TRANSVERSING THREE BODIES OF LITERATURE

The theoretical underpinnings that have shaped and guided the research reported in this book transcend three bodies of interrelated literature, viz., (1) the role of practical work in school science as perceived by different stakeholders, including curriculum designers and teachers; (2) the use of assessment as a tool to improve teaching and learning; and (3) the role of teacher beliefs and teacher thinking on their classroom actions. The reviews of these three areas in the first three sections of this appendix will be used in the fourth section as a basis to establish the rationale underlying the current study.

1. THE ROLE OF PRACTICAL WORK

The history of practical work as an integral part of school science teaching has its roots in the nineteenth century (Nott, 1997). Since then, when improvement of science teaching is considered, practical work has usually been assumed to have a special significance (Hofstein and Lunetta, 1982, 2004). In their review chapter "Research on Using Laboratory Instruction in Science" in the *Handbook of Research in Science Teaching and Learning*, Lazarowitz and Tamir (1994) offer a rationale for practical work in school science education. This rationale comprises seven dimensions and may be summarized as follows:

1. **Facilitating understanding of science concepts and theories** by providing concrete experiences to students. At the same time, practical work offers opportunities to identify students' misconceptions.

2. **Enhancing understanding of the nature of science** by involving students in actual investigations. Students may develop insights into scientific methods and expertise in using it.
3. **Developing cognitive abilities** such as problem solving, analysis, generalizing, critical thinking, applying, synthesizing, evaluating, decision-making, and creativity.
4. **Developing skills of various kinds** such as manipulative, investigative, organizational and communicative.
5. **Enhancing understanding of the concepts underlying scientific research** such as definition of a scientific problem, hypothesis, assumption, prediction, conclusion and models.
6. **Developing scientific attitudes** such as honesty, readiness to admit failure, critical assessment of the results and their limitations, curiosity, risk-taking, objectivity, precision, confidence, perseverance, responsibility, collaboration and readiness to reach consensus.
7. **Motivating students to learn science** by stimulating interest and enjoyment.

1.1 Science educators/researchers' perceptions

Despite widespread acceptance of the importance of practical work in school science curricula, controversy concerning its roles and their relative emphasis has not been resolved. Indeed, debate on this subject has existed since Todhunter's (1873) attack on the unscientific nature of school experiments and Armstrong's (1898) advocacy of the heuristic method of teaching, and has continued up to the present day (e.g. Woolnough and Allsop, 1985; Hodson, 1992, 2001). For instance, Woolnough and Allsop (1985) argued that:

> We need to deliberately and consciously separate practical work from the constraint of teaching scientific theory. We must stop using practical as a subservient strategy for teaching scientific concepts and knowledge. There are self sufficient reasons for doing practical work in science ... (p.38-9)

In contrast, Atkinson (1990) maintained that one purpose of practical work "is to promote learning of particular symbols or facts or concepts" (p.121). Kreitler and Kreitler (1974) were even more extreme. For them, practical work provides direct experiences with concepts that lead to episodes that give the concepts meaning. Practical work experiences also help students to establish the accuracy of their beliefs. However, these authors did not believe that problem solving and arousing interest and curiosity are realistic roles of practical work.

1.2 Curriculum aims and objectives

It is logical that science educators and those engaged in developing school science curricula will bring with them their own perspectives about the role of practical work. Their beliefs and thinking will eventually be reflected and manifested through the curriculum aims and objectives. As witnessed in the history of development of science curricula, there have been quite radical swings regarding the role of practical work, with several oscillations between emphasis on concept acquisition and science process skills development (e.g. Gee and Clackson, 1992; Nott, 1997).

The earliest practical work was used largely for confirming and illustrating theories (Hofstein and Lunetta, 1982). Despite Armstrong's advocacy for a more heuristic approach, the tendency to regard practical work as subservient to theory remained relatively unchanged until the 'new' science curricula of the 1960s, which began to refer to ideas like 'discovery learning'. It seemed as if these 'new' curricula had swung the pendulum towards the 'process' end of the swing. However, in reality, their main purpose was still concept acquisition, despite their good intentions to teach scientific method through practical work. This was because their 'guided-discovery' practicals were relatively tightly controlled and the 'right answer' often apparent (Gott and Duggan, 1994).

Around the same time, the 'process science movement' emerged, focusing on the cognitive processes that go through scientists' minds as they perform practical science activities. This 'process approach' was criticized (see, for example, Wellington, 1989, 1998; and Hodson, 1996) for the distorted view of science it held: that the skills and processes of science (observing, inferring, predicting and so on) could be divorced from the knowledge base; namely the laws and theories of science. Processes were to be dis-embedded from their context and content, learned and taught separately, in the hope that they could be transferable to other situations.

In most process-led science courses, practical work served to illustrate the processes in question, usually *one* at a time. This is in contrast with the latest move towards a *holistic* approach in the form of scientific investigations (Gott and Duggan, 1994) which aim to allow students to use and apply concepts, cognitive process skills and practical skills. As such, the latest development seems to lie somewhere midway between the two ends of the content-process swing.

Despite the diverse opinions and the seemingly never-ending debate, a consensus seems to be emerging among science educators and curriculum developers that we need different kinds of practical work for different purposes (Gott and Duggan, 1994). For example, *illustrative* practicals are required to demonstrate a concept already taught by the teacher; *skills* practicals aim for the development of basic manipulative skills; and *investigative* practicals aim to allow students to use concepts, cognitive processes and skills to solve a problem. Another emerging consensus seems to accept that understanding how scientific inquiry is carried out should be one of the central aims of science education (National Science Council, 1996; Duschl, 2000). That is, besides understanding the science content, students should have some understanding of the relationship between evidence and theory. In using data to support the claims they make, students need to consider the quality of the empirical evidence collected. This understanding is seen as an important component in developing an understanding of the epistemology of science and enabling students to engage in public debates involving the use and evaluation of scientific evidence (Millar and Osborne, 1998).

But do teachers think alike? Do they ever think precisely about what different sorts of practical work are for, and when, how and why they should be deployed? Or do they just let them co-exist in the muddle that has long been criticized? For example, Hodson (1988, 2001) criticized that practical work, as conducted in many schools, is ill-conceived, confused and unproductive. He opined that such practical work provides little of real educational value. In the context of investigating the extent to which student argumentation and discussion were used in inquiry-based science lessons, Newton et al. (1999), Watson and Swain (2004) echoed these concerns, namely, that the quantity and quality of discussion was low; students did not usually seek to justify claims made. These authors urged educators to clearly identify for themselves what practical work is trying to achieve, and to re-consider whether current practice is the best approach.

At this juncture, it seems appropriate to pause and ask, in the context of the current study: what are the goals of the HKAL Biology Curriculum, and which of those are thought to be achievable through practical work instruction? The following assessment objectives for practical work extracted from the HKAL Biology Syllabus (HKEA, 1996, p.154) provide an overview of the intentions behind the curriculum:

- To apply biological knowledge to solve problems and to make judgments
- To formulate working hypotheses and design tests for them, using controls where appropriate
- To observe and describe objects and phenomenon accurately
- To interpret and analyse simple biological experimental data
- To detect errors and to suggest improvements in techniques and to devise suitable experiments
- To formulate generalizations in the light of first-hand evidence

All these objectives are interwoven and contribute to the mastery of one central theme in science learning – understanding of the methods of science and the ability to apply them in problem-solving (Woolnough and Allsop, 1985). Indeed, in Hong Kong, emphasis on this theme through experimental work manifests itself through a list of more than 60 practical exercises prescribed in the A-level biology syllabus. Most teachers spend at least one-third of the curriculum time in laboratory teaching. Given that so much effort has been spent in this respect, it is both surprising and frustrating to note that most students (as well as some practising teachers), fail to master the most basic skills of investigative work, such as controlling variables, formulating hypotheses and designing experiments (Yung, 1996). The crux of the problem has been found to lie in the fact that practical activities are often conducted as highly prescriptive activities (Yip and Yung, 1998). This, in turn, may be related in part to teachers' beliefs about the nature of science and the role of practical work.

1.3 Teachers' beliefs about the nature of science and the role of practical work

As stated earlier, one of the agreed rationales for student practical work is to enhance students' understanding of the nature of science. Disappointingly, however, research has consistently shown that students' and teachers' views are not consistent with contemporary conceptions of the nature of science (Duschl, 1990; Lederman, 1992; among others). In his comprehensive review, Layton (1990) described the situation in the following way:

> The general thesis is that the philosophy of science has rarely been used in a systematic and deliberate manner as a prime source of objectives for student laboratory work. Instead, it has been resource drawn upon selectively – raided even – to underwrite purposes and practices which have their origins in considerations remote from philosophy. (p.37)

Why has this been the case? Is this due to teachers' inadequate conceptions about the nature of science? Or, are there other factors preventing teachers from teaching practical work using current philosophy of science as a prime source of objectives?

At a theoretical level, there are strong arguments pointing towards accepting, in principle, a certain relationship between a teacher's philosophical stance about the nature of science and his/her pedagogical assumptions, for instance concerning the use of laboratory work. However, this is not to imply that there is necessarily congruence between teachers' expressed views and their actual classroom practice. In fact, studies have often shown that there is *not* always a significant relationship between a teacher's beliefs about the nature of science and his/her classroom practice (e.g. Lederman and Ziedler, 1987; Brickhouse, 1989; Duschl and Wright, 1989).

Lederman (1992) suggested that the transposition of these beliefs into classroom practice is mediated by a complex set of situational variables. Indeed, the strong influences exerted by institutional and instructional constraints have as much intuitive appeal as the simplistic assumptions which guided much of the early research on the nature of science. It is quite reasonable to expect that many factors (e.g. curriculum constraints, assessment policies, student abilities and their expectations, laboratory facilities, etc.) other than a teacher's beliefs about the nature of science can influence his/her instructional approach. In reality, what actually happens in the laboratory is to a large extent governed by those aims regarded as important by teachers themselves, after taking into consideration the multitude of obstacles and demands that they are facing in their own schools (e.g. Thompson, 1975; Friedler and Tamir, 1984; Tobin, 1986; Abd-El-Khalick et al., 1998; Hodson, 1998; Lederman, 1999). That is to say, complex issues surrounding the possible influence of teachers' beliefs about the nature of science on classroom practice have yet to be resolved. It is hoped that the study reported in this book can contribute to the literature in this respect: What are the teachers' beliefs about the nature of science and the role of practical work? What consideration, if any, do the teachers give to the nature of science, amongst many other factors, in their teaching of practical work? Which is more influential, constraints or ideology? And why?

2. ASSESSMENT REFORM

This section aims to explore why and how assessment reforms can/cannot work to influence teaching and learning. It begins with tracing the rationale behind the use of public examinations and their power to change teaching and learning, firstly in the context of the educational setting in general, and then specifically in the context of practical science teaching.

2.1 Assessment reform as a tool to improve teaching and learning

As far back as 1877, Latham (1877, p.1) characterised examination as an "encroaching power" that was influencing education. More recently, assessment has been identified as a key mechanism for monitoring and intervening in the educational process with an emphasis on improving teaching and learning, both around the world (Torrance, 1995a; Hamilton,

2003) and locally in Hong Kong (ROPES, 1998). Innovations implemented in this context in the UK have been referred to as assessment-led curriculum development (Gipps, 1986; Grant, 1989), in the USA. as measurement-driven instruction (Popham, 1987). In Popham's terms, assessment directs teachers' attention to the content of test items, acting as powerful "curricular magnets" (p.680). In high-stakes assessments, in which the results of mandated tests trigger rewards, sanctions, or public scrutiny and loss of professional status, teachers will be motivated to pursue the objectives that the test embodies. All these reflect a trend towards a reliance on using assessment to influence policy and instruction to such an extent that Hamilton and Koretz (2002) characterize the current policy context in the USA as a test-based accountability system.

The theoretical perspectives underpinning such reforms and changes derive from a variety of arguments which relate to issues of curriculum quality and curriculum development, and to questions of how children learn. For example, it is argued that the quality of teaching will be improved if higher-order skills and competencies such as problem-solving, investigation and analysis are included in what is assessed (e.g. Resnick and Resnick, 1992; Madaus, 1993). The HKAL Biology TAS, which is the focal context for the current study, drew upon the above theories for its conceptualization and formulation, employing a problem-solving and investigative approach to test students (HKEA, 1997).

Based on evidence drawn from the UK, Torrance (1995b) warned against the danger of these optimistic views about the potentially positive impact of assessment on teaching. He criticised them as reflecting a very simplistic and mechanistic view of the implementation process; that is, improve the test and you will improve teaching. Torrance criticised such an idea as linear and ideal. He argued that "just as poor assessment practices can impact negatively on teaching and learning, it does not automatically follow that better assessment practices will impact positively". (p.46)

Hamilton (2003) alluded to the same point in her review of studies conducted in the USA. Locally, in Hong Kong, Andrew (1994) also commented that such a reform strategy – using assessment as a tool – is a "blunt instrument" for bringing about pedagogical changes. He asserted that the actual teaching and learning situation is clearly far more complex than proponents of alternative assessment suggest. Each different educational context (school environment, messages from administration, expectations of other teachers, students) plays a key role in facilitating or detracting from the possibility of change.

The present study is situated within this intense backdrop of debate. Hong Kong is strongly influenced by the overall world-wide assessment paradigm shift in education (Gipps, 1994). The local public examinations administering body, HKEA, is very conscious of the backwash effect of the public examinations on classroom teaching, and much thought has been given to how the examination process can be used to bring about positive and constructive change to the system (HKEA, 1994, p.80). With this purpose in mind, many innovations have been launched in recent years, including the HKAL Biology TAS – the context of the present study.

The HKEA hoped that the flexibility provided by the TAS would have a liberating influence on the biology curriculum, and would bring about a host of desirable changes, such as achievement of a wider range of objectives associated with practical work, facilitating a diversity of practical work for the students, and so on (Pang, 1992; Yung, 1992). It was also hoped that teachers would pay more attention to the formative use of assessment information to improve teaching and learning – a goal that converges with Black and Wiliam's (1998) recommendation in their comprehensive review of studies on the use of formative assessment to improve student learning. However, as Hargreaves (1989) has pointed out, assessment reforms bring no automatic pedagogical shifts, only the creation of opportunity for alternatives to be explored more thoroughly. Gipps (1994) added that the educational assessment paradigm depends upon a high level of understanding and involvement by teachers in assessment practices. Indeed, there is evidence that student-assessment is an area of difficulty for both teacher trainees (Booth, 1993; Cizek, 2000) and many experienced teachers (Butterfield, 1993; Khattri et al., 1998; Butterfield at al., 1999; Black et al., 2003). This constitutes a major source of problems for the implementation of the new forms of assessment.

2.2 Problems in implementing assessment reforms

In reviewing the conditions under which assessment reforms might have the greatest impact on classroom teaching, Airasian (1988) drew attention to the practical problem of whether or not teachers are willing and able to implement the new subject matter and teaching methods. Drawing on evidence from a number of assessment reform initiatives in the UK, Torrance (1995b) also emphasised the importance of teachers' understanding of the curricular and the pedagogic implications of their becoming involved in new approaches of assessment in order that their anticipated benefits can be maximized.

In fact, findings from different studies on the effect of assessment reforms on classroom teaching have been inconsistent and somewhat context-specific (Khattri et al., 1998; Firestone et al., 2004). In some cases, teachers were reportedly encouraged to use a wider variety of teaching methods (Buchan, 1993). In other cases, however, teachers were very much bothered by having to assume the dual roles of assessors and teachers (Donnelly et al., 1994; Paechter, 1995). In the worst situations, the formative functions of assessment were supplanted by the summative function. Teachers became reluctant to help students (Torrance, 1991; Jones et al., 1992; Broadfoot, 1995); teaching and assessment remained largely polarized (Peterson, 1992); assessment did not arise as a natural consequence of teaching, nor had the outcomes been used to inform the process of teaching (Buchan, 1993).

The difficulties that teachers have with assessment, apart from other factors, may lie in their inherent conceptualisation of assessment (Butterfield et al., 1999; Black et al., 2003). It is possible that concepts that are presented as unproblematic in official documents are, in fact, insufficiently theorised and conceptualised for teachers to be confident in their use in schools. In such cases, changes in assessment *per se* would then run the risk of being interpreted by the teachers within a traditional testing paradigm rather than a pedagogical paradigm (Torrance, 1995b). When that is the case, there are likely to be queries and anxieties about the subjectivity of teacher marking, the criteria to be used, how grades would be made comparable across schools, and so on. Torrance highlighted the importance of alerting teachers to, and discussing with them, how the changes might broaden the curriculum and improve the process of teaching and learning. He stressed that simply involving teachers in the assessment process did not automatically stimulate such discussion among them. Further development of new assessment initiatives should be undertaken, with full regard to problems of teacher interpretation and mediation at school level, and with considerably more attention to the literature of educational innovation and the management of change.

In a parallel vein, basing his claim on experience in the USA., Hamilton (2003) pointed out that the influence of testing on classroom instruction will in most cases "interact with other factors such as teachers' own beliefs and knowledge about pedagogy and subject matter, their professional development experiences, their access to necessary resources including appropriate curriculum and instructional materials" (p.33). In other words, educational reform efforts directed towards instructional improvement should acknowledge the challenges presented by the requisite conceptual changes on the part of the teachers. Changes of this type are not simply brought about by the imposition of new teaching materials. Conceptual changes are seldom achieved without attending to the beliefs of those who are targets of change – teachers – and the conditions of the educational environments in which they function. A key challenge within this context is that teachers are now required to perform the dual roles of teachers and assessors.

2.3 Practical work assessment reforms

This section will focus specifically on reforms related to the assessment of practical work. It aims to explore the intentions behind the reforms and, in particular, to find out if there is any reason that is specific to the nature of work that is to be assessed.

2.3.1 Towards more school-based assessment

The presumed centrality of students' practical work in the science curriculum, reinforced by a variety of curriculum initiatives in the 1960s, has raised a number of questions about how such work might best be assessed and accredited (Buchan et al., 1992). The conventional external practical examination has come to be seen as seriously and manifestly defective on technical grounds as well as in terms of its costliness. Caillods et al. (1996) have found that the expense of practical examination arises partly from the need to assess 'under examination conditions' in order to provide a fair test. This requires that standard apparatus and consumables should be available. If specimens of any kind are involved, this may add to the problems, since large numbers may have to be procured, quality assured, and distribution problems solved. Additional costs will be involved in providing time for school laboratories to made available for examinations, which often disrupts normal teaching for substantial periods. A further cost arises if any element of performance, as opposed to outcome, is to be assessed since this can only be undertaken as it happens. Elements of technique, some aspects of observational skills, and design and experimental skills can only be properly assessed by on-the-spot observations, following the processes a candidate goes through. This is both time-consuming and costly.

Worse still, is the backwash effect of the practical examinations on the curriculum, as evidenced in relation to the HKAL Biology curriculum in the past. Due to the large number of candidates, several comparable papers had to be set for the HKAL Practical Examination each year. This, together with the technical difficulties associated with setting up the practical examinations, greatly restricted the kind of experiments and techniques that could be included in the examination. As a result, teachers tended to ensure that their students developed a high proficiency in the restricted range of experiments and techniques that were commonly examined in the practical examination. A letter of complaint from a teacher to the HKEA provides a glimpse of the extent to which the curriculum had been distorted, as well as how ill-conceived and unproductive the teaching of practical work had become:

> (The students) normally dissect about 20 rats each before sitting for the A-level practical examination. Diligent students may dissect as many as 40 rats... (Yung, 1992, p.92)

There was clear recognition that the format of the practical examination had adversely influenced the laboratory work that was being done (Yung, 1992). Thus, the subsequent emergence of a compulsory component of school-based assessment of practical competence in science can be seen as a response to the need to assess such competencies within a system that accommodates a large candidature. It can also be regarded as a method introduced to address the backwash effect of the practical assessment on the curriculum.

In analysing the relationship between the science curriculum and its assessment in different countries, Caillods et al. (1996) conclude that the kind of outcomes that science curricula often aspire to can only generally be assessed through a combination of written tasks under examination conditions and the performance of practically based activities that may extend over a relatively long time and involve collaborative work with others. Thus, many outcomes cannot be assessed unless some elements of school-based assessment are involved.

School-based assessment is common in many school systems in developed countries. In her review of practices of different countries, Pennycuick (1990) drew attention to the range of approaches used and some of the problems that have arisen. Prominent amongst them are the increased assessment burden on teachers, reduced reliability, more sophisticated moderation, and the prevention of collusion, cheating and favouritism. Caillods et al. (1996) thus warned that adoption of school-based assessment should only occur when the following conditions are met: an adequate educational infrastructure; sufficient training of science teachers in assessment techniques; and public acceptability of school-based assessment. These are especially important if the assessment has selection consequences, as is still the case for Hong Kong (Biggs, 1995).

2.3.2 Towards more investigative practical work

Buchan et al. (1992) noted that the shift from external practical examination to school-based assessment of students' practical competence in the UK was accompanied by a number of parallel shifts in the rhetoric of practical work in school science education. Such shifts stemmed from a change in perceptions of what practical work in science was *for,* and of the relationship that had hitherto existed between teaching and learning on the one hand, and assessment on the other. The kind of practical work undertaken in schools reflected the need for teachers to prepare their students for further studies in the universities. Thus, chemistry students were required to develop competence in qualitative and volumetric analysis. Biology students focused on dissection, anatomy, morphology and histology, in order to better prepare themselves for medical education. Physics students, on the other hand, spent much time verifying laws and re-determining physical constants. During the 1960s, there were attempts to redirect school laboratory teaching towards open-ended, 'original' problem-solving and experimental activities. However, ideas about the role of practical work in school science varied. The prominent concern of some was the use of practical work for the development of concepts and for training in inquiry skills (Lock, 1986), while others were only concerned with the acquisition of practical techniques deemed important to the future scientist.

In the 1980s, the 'process science movement' was revived in England with the establishment of the Assessment Performance Unit. The emphasis was on introducing students to the methods of science. Buchan et al. (1992) commented that such an emphasis, coupled with that on the assessment of the practical components of 'science process', had more than an element of the realization of the long-standing assumption that school science education should consist essentially of teaching children to "behave in the way practising scientists behave" (p.337). However, some process-led courses had been criticized for their content because the concepts they employed to focus on different processes often lacked continuity. Practical work served to illustrate the process in question in isolation. This is in contrast with the latest move towards a *holistic* approach in the form of 'scientific investigations', which aim to provide students with opportunities to use and apply concepts, cognitive process skills and practical skills to solve scientific problems. This latter aim underlies also the HKAL Biology TAS – the context of the current study (Yip and Yung, 1998).

2.3.3 Towards a more holistic approach

Parallel to the debate about developing students' process skills 'one at a time' versus that of using a holistic approach in the form of 'scientific investigation' is the debate about using an 'atomistic' approach or a 'holistic' approach to the assessment of practical competence. At the core of the skills-based approach to assessment, and consequently to teaching, is a commitment to the notion that scientific method comprises a set of discrete, context-independent, generalisable and transferable processes – involving observing, classifying, measuring, hypothesising, and so on (e.g. Bryce et al., 1983). Thus, doing science (carrying out scientific inquiries and investigations, and solving scientific problems) consists of the application of these skills – learned elsewhere – in a particular context. Indeed, some assessment schemes in science (e.g. TAPS1 – Bryce et al., 1987) spell this out quite clearly in their advocacy of such a 'step-up model' for teaching scientific inquiry.

Based on the theory-impregnated nature of scientific processes, Hodson (1993) challenged this skills-based approach by saying that:

> Science is more than the sum of its parts... It follows that students cannot acquire expertise in doing science by acquiring a capacity to carry out a series of unrelated exercises. In other words, skills-based approaches are based on a false model of science. (p.129)

As such, he argued that a skills-based approach to assessment does not constitute a valid representation of doing science, and hence cannot effectively serve the summative function. In Hodson's view, assessment schemes should only be deemed satisfactory if they fulfill four basic functions. First, a summative function: they should provide some description of a student's level of attainment in all aspects of the course, at the end of that course. Second, a formative function: they should enable teachers to monitor learning progress in order to plan more effectively for the further learning of each student. Third, an evaluative function: they should provide teachers with information about the effectiveness of the curriculum experiences provided, in order to assist them in curriculum decision-making and course planning. Fourth, an educative function: they should enhance and promote learning by engaging students in interesting, challenging and significant experiences aimed at developing further insights and understanding. Educative assessment requires that assessment become part of learning, not something additional to it. Engaging in the assessment activities should, in itself, be a significant learning opportunity and experience.

According to Hodson, in the context of assessing students' practical work, in order to have a summative function, assessment activities must be based on a philosophically sound model of science. In other words, the activities on which assessment is based must collectively constitute a valid representation of doing science. In order to have a formative function, the approach must be capable of providing information helpful to teachers in diagnosing current learning for all students, recognising what constitutes progress in doing science, and devising appropriate teaching and learning strategies. In order to have an evaluative function, the approach must provide information about the successes and failures of the learning enterprise. Thus, it must embody a clear understanding of what constitutes the goal (in this case, doing science), and must be effective in helping teachers to identify any deficiencies in their teaching approach. To have an educative function, assessment activities must encourage and involve further learning.

With regard to the formative, evaluative and educative functions, Hodson (1993) asserted that skills-based approaches to assessment would lead to trivialisation and bad educational practice. In his words,

> Skills-based approaches to assessment inhibit good educational practice by directing attention away from investigational laboratory work toward sterile, de-contextualised, disconnected practical exercises conducted under strict teacher control... Skills-based methods trivialise through their relentless pursuit of a spurious and distracting reliability... Because assessment is restricted to the measurable there is a danger that only that which is measurable is taught. Ease, precision and reliability of measurement determines the curriculum. (p.129-130)

Woolnough and Toh (1990) raised a similar concern that the skills-based approach would reinforce an unproven method of learning to do science investigations, whereby whole investigations are replaced by easily managed, easily assessed practical tasks and exercises.

In contrast, holistic assessment is seen to have a powerful educative function, in that it requires students to engage in activities which contribute directly to the further development of their understanding of what doing science involves and to their capacity to engage in successful scientific investigations, thereby reaching a major goal in the quest for scientific literacy (Hodson, 1992).

In all, with their implied concern for engaging students in scientific enquiry, skills-based approaches are seen by some educators to be guilty of a fundamental misunderstanding of the purpose of practical work. These educators assert that practical skills have little or no value in themselves though they may be necessary in order for students to engage in one of the real purposes of practical work: doing science for themselves and by themselves, and gaining some expertise, understanding and enjoyment from it. Thus they think that to focus assessment on basic skills is to miss the point. Reliability may have been gained, but at the expense of validity because this no longer constitutes doing science. As a consequence, skills-based assessments are no longer valid from a summative, formative or evaluative perspective, and are counter-productive from an educative point of view.

But to what extent do teachers share such views? And how do their perceptions of these issues relate to their classroom practices and teaching approaches? There is UK-based evidence suggesting that teachers' classroom actions may somehow be influenced by the assessment framework of some school-based schemes for assessment of practical competence. For example, with respect to the evaluative function, Buchan (1993) reported evidence of teachers making use of the assessment framework of the assessment schemes to improve their teaching. In her words:

> The introduction of the assessment schemes has provided a framework for practical science which teachers have used to analyze the practical experiences they provide for students. They therefore describe (themselves) being able to provide more meaningful activities (for the students). (p.13)

With respect to the educative function, Buchan (1993) was also able to identify innovative tasks in some classrooms. Nonetheless, problems have also been found with the school-based assessment schemes of practical competence, one of which is associated with the underlying notion of fairness (Donnelly et al., 1994). This is particularly influential in governing teachers' attitudes towards their role as internal assessors. There are tensions evident in the extent to which teachers are willing to help students whilst carrying out an assessment. Buchan (1993) reported that:

> Most commonly, students work under what can be described as 'strict examination conditions'... [And there are] intense dislikes, by many [students], of being required to work in isolation in the laboratory. Practical situations evoke a need to test out their ideas on others, prior to implementation, and this is denied them during practical assessments. (p.11-14)

That is to say, the formative, evaluative and educative functions of assessment become supplanted by the summative function. Teachers become reluctant to help students. The practice of working in pairs or small groups – a common feature of real scientific research and good pedagogy – is discouraged. This means that "teaching and assessment remain largely polarized... Assessment has not arisen as a natural consequence of teaching and nor have the outcomes been used to inform the process of teaching" – a rather grim scenario observed by Buchan (1993) in her study.

To sum up, there are both harmful and beneficial effects of implementing school-based continuous assessment schemes. Whether these schemes are implemented in a manner that helps to achieve Hodson's four assessment functions depends very much on (1) teachers' understanding of these aims, in particular, those of the formative, evaluative and educative functions, and (2) how much they ascribe to these aims. That is, we need to ask the question: What are teachers' beliefs about the purposes and functions of assessing students' practical work? In other words, what are teachers' understanding/perceptions of the TAS? In addition, we need to know, in much finer detail, how assessment of practical work is actually carried out at the classroom level and what difference it makes, if any, to the culture of classroom life. In turn, we might then be able to offer descriptions of how the TAS has actually impacted on the learning process and under what conditions it might realise some of its more ambitious claims, including its formative, evaluative and educative functions.

3. TEACHERS' BELIEFS

This section aims to review existing literature on the role played by the science teacher's beliefs and aspirations in shaping his/her own classroom practices.

3.1 Teachers' beliefs and their practices

In the previous sections, teaching involving practical work and its assessment has been depicted as principles or rationales at the conceptual level, and as curriculum guidelines and assessment regulations at a more concrete level. These depictions also represent the shaping of the teaching of practical work by different people or organizations. But what form and shape the teaching of practical work will take in the school laboratory, and how it will be experienced by the students, will ultimately depend on the shaping by their teachers.

The importance of teachers in curriculum implementation has long been recognized (Connelly and Ben-Peretz, 1980; Olson, 1981); but the prevailing view of findings of curriculum research has been a deficit view, with teacher practice in the classroom falling short of the innovator's intentions (Olson, 1981; Cronin-Jones, 1991). Studies of the role of the teacher in curriculum implementation document the extent of teachers' unwillingness to re-orient, but say less about why this happens and what should be done about it. They give only glimpses of the underlying concerns of teachers that appear to have influenced their use of the new ideas.

In the research carried out by Olson (1981) and Mitchener and Anderson (1989), which investigated how teachers implemented new initiatives in the curriculum, it was revealed that when the philosophy underlying a curricular innovation differed significantly from that held by a teacher, the teacher had to re-structure his/her beliefs, or to "domesticate" the curriculum in order to fit it into his/her belief system. Such re-structuring or domestication may be problematic or even "personally threatening".

Such results from curriculum studies on teacher thinking call for significant attention to be given to teachers' beliefs, in order that innovations such as the teaching and learning activities advocated for the TAS can have any chance of an enthusiastic, thorough and sustainable implementation in the laboratory. The TAS demands a new definition or re-formulation of the role of the teacher – a dual role of assessor and teacher. The change over from a practical examination to a school-based continuous assessment scheme also demands that teachers have a different conception of assessment and its relationship to learning. Philosophically, the different positions imply different conceptions of what practical work teaching is, different beliefs about the kind of practical work that is worth doing, what practical work is for, and different views of teaching, learning and assessment. So much so that these entail a shift in thinking which is referred to by Gipps (1994) as a paradigm shift – from a "testing" model to a broader model of "educational assessment". Whether these new beliefs can be translated into pedagogical form depends on the personal beliefs of the teachers, on their ability to organize the changes needed, and on their preparedness to meet these challenges.

3.2 Teachers' coping mechanisms and their structure of awareness

Early studies on the teaching of practical work in schools (e.g. Kerr, 1963; Thompson, 1975) tended to adopt a reductionist approach in which a list of teaching objectives for practical work was set *a priori* and teachers were asked to rate their relative importance. Data was analyzed quantitatively and no attention was given to individuality and variation due to differences in local context. Later studies, (e.g. Tobin, 1986) realizing the complexity of the problem, have adopted a more qualitative and naturalistic approach to the problem and involve greater consideration of the teacher's perspective. As a result, the kinds of teaching-learning activities that are actually occurring inside the school laboratory are illuminated, as are the various confounding variables that affect laboratory teaching.

Qualitative studies of this kind are definitely more informative than the previous ones. However, they still fall short of explaining how the teacher resolves all the constraints, problems and dilemmas, and what are the justifications and concerns behind their pedagogical decisions. Some studies (e.g. Brickhouse and Bodner, 1992; Fischler, 1994; Blake, Jr. 2002; Roehrig et al., 2004; Verjovsky and Waldegg, 2005) have begun to look into the mechanisms teachers create to circumvent the constraints on their teaching or how they adjust their teaching to fit their circumstances. Various labels have been assigned to these decision-making and/or coping mechanisms, such as the functional paradigm (Lantz and Kass, 1987), models of reality (Duschl and Wright, 1989), theory-in-use (Benson, 1989), teacher mind frames (Tobin et al., 1990), teacher practical knowledge (Duffee and Aikenhead, 1992), cultural myths (Tobin and McRobbie, 1996), personal theories (Richardson, 1996), practical theories (Hillock, 1999) and the structure of teachers' awareness (Marton, 1994).

In common with many other qualitative studies, the study described in this book has been influenced by different perspectives at different times of its development, especially during its data interpretation stage. Overall, its interpretive framework has been predominately influenced by Marton's experiential framework. It must be emphasized that the interpretive framework that was drawn upon in analyses was chosen because it appeared to be highly consonant with the data. Below is a brief description of the main features of that framework.

Most of the studies mentioned above put emphasis on:

- explaining the teachers' behaviours or acts in terms of their mental representations, e.g. personal theories, practical theories, theory-in-use, practical knowledge, or
- explaining the teachers' consciousness in terms of the influence of the sociocultural milieu on them e.g. cultural myths, functional paradigm, models of reality, teacher mind frames, etc.

In Marton and Booth's terms (1997), studies in the first category above are trying to explain the "outer" (the teachers' acts) in terms of the "inner" (the teachers' mental representations), whereas the second group of studies views teachers' consciousness as the "inner" and the social, cultural world as the "outer". In other words, there are two opposing schools of thoughts among studies of relationships between teachers' acts and their thinking, their focuses being on different sides of the borderline between the "inner" and the "outer". According to Marton and Booth (1997), however, researchers should not consider "person" and "world" as being separate. They assert that researchers should not resort to hypothetical mental structures divorced from the world. Nor should researchers resort to the social, cultural world as seen by the researcher only. These authors stress that "people live in a world which they – and not only the researchers – experience. They are affected by what affects *them*, and not by what affects the researchers" (p.13) (original emphasis).

In brief, Marton and Booth call for taking the experiences of people seriously and exploring the physical, the social and the cultural world they experience. The world which the researchers deal with should be *the world as experienced by people* (in our case, the teachers) – neither individual constructions nor independent realities. These authors also argue that the people (in our case, the teachers) we deal with are people experiencing aspects of the world – they are neither bearers of mental structures nor behaviorist actors. Of course, the researcher does not have access to the experiential world of the teacher directly, but has to interpret what the teacher says and why the teacher behaves in a certain way.

Adopting this line of argument, the dividing line between the "outer"and the "inner" disappears in the present study. There are not two things – the "outer" and the "inner"; and one is not held to explain the other. There is not a world "out there" and a subjective world "in here". The world is not constructed by the teacher, nor is it imposed upon him/her; it is *constituted* as an internal relation between them. In other words, the way in which a teacher handles a certain situation to some extent *reflects* the teacher's understanding of the phenomena involved, and to some extent *changes* his/her understanding of the phenomena involved. That is to say, a teacher's understanding of the situation itself and his/her understanding of the phenomena that are relevant to understanding the situation are seen as related.

Marton and Booth (1997) crystallize all these thoughts into an explanatory framework called the *nature of awareness* (p.82) which, amongst others, encompasses the following main features:

- Certain structures of awareness are implied by certain ways of understanding
- The person is simultaneously aware of certain aspects of a situation or a phenomenon[1]
- The person's awareness of certain aspects implies a tacit awareness of other aspects
- Certain aspects become figural, in focus or focal, whereas other aspects recede to the background.

Applying this explanatory framework in the world of teaching as experienced by the teachers, Marton (1994) described the structure of teachers' awareness in the following way:

> There are various aspects of teaching, of aims of teaching, of the particular content, in relation to which teachers differ significantly. These aspects are different layers of the teachers' awareness, which are

[1] According to Marton and Booth (1997), a *situation* is always experienced with a context, a time, and a place, whereas a *phenomenon* is experienced as abstracted from or transcending such anchorage. In most situations, however, the two are inextricably intertwined. That is, nobody can experience a phenomenon in the absence of a situation. In their own words, the proposition that "a situation can be experienced only in terms of that which transcends it follows from our ability to make sense of the here and now only through the experiences which precede it: The here and now is experienced in terms which link it with the past and reflect experienced similarities, differences, or both. We refer to the wholeness of what we experience to be simultaneously present as a *situation,* whereas we call entities that transcend the situation, which link it with other situations and lend meaning to it, *phenomena*." (p.82-83)

> simultaneously present. In the concrete way of teaching a concrete content, all the increasingly general aspects of the teachers' awareness are inherent. Everything is there - admittedly not always clearly; not always explicitly - in every moment of the very teaching of the very content. (p.41)

In other words, of the various aspects of teaching, only certain relevant aspects will be taken into consideration by the teacher in any specific instance of teaching, and these will represent the directedness of the teacher's intentionality.

Of interest to the present study is the teachers' awareness of the aims of education in general, and of the aims of science education, practical work instruction, school-based assessment of practical work in the specific act of teaching practical work inside their classrooms. In particular, among the various general aspects of teachers' awareness, "pressures of accountability" is commonly reported by many studies (e.g. Duschl and Wright, 1989; Brickhouse and Bodner, 1992) as a constraint affecting teachers' pedagogical decisions. Given the situation in Hong Kong, where the emphasis on examination for selection purposes is still much stronger than in other places (Biggs, 1996), it is anticipated that this and other related concerns like "to help students pass the examination" and "to maintain student respect" are bound to be focal in local teachers' structure of awareness. Is this really the case? In what ways and to what extent does such awareness affect the teachers' practices in their classrooms?

3.3 Teachers' aspirations and their classroom actions

The current research was set up mainly to investigate if there exist relationships among (1) secondary science teachers' classroom actions, (2) their beliefs about science, teaching and learning, and (3) their understanding of the school-based assessment reform in practical science (i.e. the TAS). In order to make the answers to these questions meaningful, it was necessary to embed the answers in the contexts in which they existed. I adopted the position that valid interpretation of belief-practice linkages can only occur in conjunction with an exploration of the other aspects and factors that teachers may draw into their focal awareness, such as curriculum aims, student expectations, class size, available material resources. (e.g. Shavelson and Stern, 1981; Duschl and Wright, 1989). Nevertheless, I chose to focus on four central tenets: beliefs, goals, behaviour, and the context of action (Tobin and McRobbie, 1996), while bearing in mind that there could always be other aspects in the teachers' focal awareness that might not be included within these four tenets. It was hoped that these four mutually interactive elements would comprise a framework that includes the dialectical interaction between the individual teachers and the sociocultural milieu in which they enacted their professional lives.

Such a theoretical stance echoes Nespor's (1987) suggestion that more attention be paid to the *goals* pursued by teachers (which may be multiple, conflicting, and not at all related to optimizing student learning) and to their subjective interpretations of classroom processes. Nespor substantiated her recommendation by referring to two teachers in her study. One teacher saw teaching mainly as a job, a way of making a living. The other teacher, by contrast, looked at teaching as a sort of moral mission to socialize children and improve the community; money for her was secondary. In brief, teaching took on completely different meanings for these two teachers. Failure to recognize this would spoil any attempt to make sense of what the teachers did in their classroom or why they did it. Accordingly, she believed that "to understand teaching from teachers' perspectives we have to understand the beliefs with which *they* define their work" (my emphasis) (Nespor, 1987, p.323). She believes that classroom structures have their sources in teachers' beliefs and are maintained over time only as a result of the *goal-directed* actions of teachers (and students). Louden (1991) echoed the importance of such an approach in relation to understanding teaching; as follows.

> [T]eaching is a struggle to discover and maintain a settled practice, a set of routines and patterns of action which resolve the problems posed by particular subjects and groups of children. These patterns, content and resolutions to familiar classroom problems are shaped by each teacher's biography and professional experience. The meaning of these patterns of action only becomes clear when they are set in the context of a teacher's personal and professional history, *her hopes and dreams for teaching*, and the school in which she works. (p. xi) (my emphasis)

From the above line of argument, in order to identify what goals an individual teacher aspires to, it was necessary for me to probe into the teacher's biography or personal experiences that may have influenced his/her classroom action. Of relevance to the current study, are those related to the teachers' beliefs about the nature of science (e.g. Brickhouse, 1989; Tobin and McRobbie, 1997), about science teaching and learning (e.g. Hewson and Hewson, 1989), and about assessment (e.g. Briscoe, 1993; Butterfield et al., 1999). Teachers' beliefs about the nature of science, for instance, would be embedded in the teachers' attitudes and backgrounds in science and science education. To uncover and place these beliefs in the appropriate contextual perspective would necessitate at least an exploration of the effects of the teachers' experiences both as a student (in the past) and as a practitioner. The portraits of individual teachers reported at the beginning of each case story in Chapters 4 to 11 serve this purpose.

4. ESTABLISHING THE RATIONALE UNDERLYING THE CURRENT STUDY

This appendix has reviewed the role of practical work in the school science curriculum as perceived by different stakeholders, including curriculum designers and teachers. Discrepancies were observed between the intended outcomes and what was observed in classrooms. Problems were identified and teachers' concerns highlighted. Attempts to improve the teaching-learning situation using assessment as a tool were also reviewed. However, one notable feature from the review of these two bodies of literature is that most early studies tended to look at the problem(s) at a '*system*' or collective level. Seldom did they focus on individual teachers or look at the problem from the teacher's perspective. Even if they did, as in studies of the relationship between a teacher's belief in the nature of science and his/her teaching, the variables under study were usually singled out and studied one at a time. Seldom did any of the studies reviewed probe deeply into the teachers' beliefs in a holistic manner, as recommended by the literature reviewed in the third part of this appendix – studies on the relationship between teachers' beliefs and their practices. In addition, the majority of early studies tended to resort to hypothetical mental structures that are divorced from the world (i.e. there is a separation between the "inner" and the "outer" aspects of a person) as the explanatory framework for relationships between the teachers' acts and thinking. This is in contrast to Marton and Booth's advocated method of seriously considering the experiences of people and exploring the physical, the social and the cultural world they experience. In their words, the researcher should deal with the world as experienced by the teachers and *not* the teachers as bearers of mental structures or behavourist actors. The teacher's understanding of a certain situation itself and his/her understanding of the phenomena that are relevant to understanding the situation are somehow related and drawn into the teacher's focal awareness, and these represent the directedness of the teacher's intentionality. It is this idea of the structure of teachers' awareness that guided the present investigation into the possible relationships among: (1) the classroom actions of secondary science teachers in the TAS; (2) secondary science teachers' beliefs about science, teaching and learning; and (3) their understanding / perceptions of the school-based assessment reform in practical science (i.e. the TAS).

In all, the rationale for the study reported in this book was that in order to understand how change might be initiated and sustained, it is first necessary to understand why teachers do what they do, through investigations of the world as *experienced* by them, including their beliefs and their classroom actions.

Appendix B

REFLECTING ON THE RESEARCH METHODOLOGY

An outline of the research methodology was given in Chapter 1. To cater for interested readers, this appendix provides a more comprehensive account of the research approach and methodology adopted. It culminates with a reflection on some of the methodological issues.

1. A CURRICULUM ENACTMENT PERSPECTIVE

In studying how practical work was taught and assessed in schools in the context of the TAS, the particular focus and interest of this research was on how the curriculum in practice was being shaped by the teachers. As such, the biology teacher was regarded as integral to the implementation of the curriculum under the influence of the TAS. Thus, this study was basically a research on curriculum implementation with an enactment perspective, as distinct from the "fidelity" perspective and the "mutual adaptation" perspective. According to Synder et al. (1992), a curriculum enactment perspective focuses on:

> how a curriculum takes form through the evolving constructions of teachers and students ... (T)he interest is not in measuring how faithfully the curriculum is carried out (the fidelity perspective), nor even so much how the original curriculum is adapted (the mutual adaptation perspective), but rather in describing and understanding the meaning given to the evolving curriculum by those creating it – the teacher and students. (p.430)

In the present study, the focus was on the teachers' meaning-perspective of their own actions in the classroom. This was manifested through asking the research question: What are teachers' perceptions / understanding of the TAS? In other words, the prime concern was to describe and understand the meaning given to the evolving TAS by the teachers concerned.

2. AN INTERPRETIVE APPROACH

The research approach taken in this study can be classified as the interpretive approach (Erickson, 1986), where the prime aim was to elucidate and interpret the meaning-perspectives of the teacher in making curricular choices and decisions in the context of the TAS. This involved careful observation and recording of classroom events over an extended period of time, collection and analysis of documentary evidence, and in-depth interviews with teachers. According to Erickson (1986), this interpretive approach is appropriate when one wants to know more about, among others, "the specific structure of occurrences rather than their general character and overall distribution" and "the meaning-perspectives of the particular actors in the particular events" (p.121). In the current study, the focus was on the teachers' meaning-perspectives of their own actions in the classroom with respect to what the TAS meant for them.

Within this interpretive approach, case studies were used to study and report the teachers' classroom actions and underlying beliefs. According to Shulman (1986), cases are not only thick descriptions of concrete details of events or incidents. The data contained in the case studies have to be "explicated, interpreted, argued, dissected and reassembled" (p.12), and the knowledge generated from case studies has to rely on the "conceptual apparatus" of the researcher (p.11). The interpretive framework guiding this research in its initial stages consisted of four conceptual constructs derived from a review of the relevant literature reported in Appendix A. These constructs were related to the teachers'

- beliefs about the nature of science
- beliefs about teaching and learning
- beliefs about the role of practical work in school science education
- personal aspirations and visions for education

The first two of these, 'beliefs about the nature of science', and 'beliefs about teaching and learning' were based on Hewson and Hewson's (1988) notion of "an appropriate Conception of Teaching Science" (CTS). The notion of CTS was, in turn, derived from a review of the relevant literature, in particular, on how students can learn science effectively. Based on this notion, Hewson and Hewson (1989) developed an interview task which can be used to probe teachers' conceptions of teaching science. This was also the main instrument used in the current study to probe the teachers' beliefs about the nature of science as well as their beliefs about teaching and learning.

In line with the focus of the present study on the teaching of practical work, a third construct – 'role of practical work' – was added to the interpretive framework. The components of this construct were largely derived from a review of the relevant literature, including Lazarowitz and Tamir's (1994) review article on "*Research on Using Laboratory Instruction in Science*". The fourth construct – teachers' 'personal aspiration and vision of education' – was added to the interpretive framework in response to the ample evidence pointing to the need to attend to the goals-directed actions of teachers (Nespor, 1987).

However, as the data emerged, it appeared that the conceptual constructs formulated at the outset of the research were not sufficient in explicating all the data collected. As reported in earlier chapters, two other conceptual constructs – teachers' views of fairness and their sense of professionalism – formed powerful "conceptual apparatuses" which have helped me to make sense of *all* the data collected across all the cases.

Thus, the four constructs developed prior to the study were *a priori* theoretical categories guiding the research at its initial stages. Having been immersed in and familiarized with the data, the enactment perspective allowed me to recognize two new constructs based on abstractions from the data collected. Such an evolution and development of new constructs from *a priori* theoretical constructs during the process of enquiry parallels the distinction made by Gudmunsdottir (1991) between the "emic" and "etic" perspectives, which she borrowed from anthropological studies:

> The emic view represents the natives' perspective and the etic view constitutes the researchers' theoretical categories, and in qualitative research researchers need both. Researchers need the etic perspective (through the theoretical categories) to approach the phenomena he/she studies, to know what to look for. In this way, the etic perspective is a kind of vehicle to access the emic perspective. Once researchers have approached the emic perspective, they know what to look for; they have to let go of the etic perspectives and try to understand the emic view on its own terms. Once this is achieved, it is appropriate to bring back the theoretical categories to explain the emic view. (p.415)

To sum up, case studies were used in the current study as a means of reporting and highlighting the teachers' classroom actions in teaching practical work, and their underlying beliefs.

3. MAJOR CONSIDERATIONS IN DRAWING UP THE RESEARCH DESIGN

Since the present research had its focus primarily on the possible relationship between the teachers' acts and their beliefs, two major categories of data were collected – the teachers' classroom practices in the TAS and the beliefs underlying their practices. The present research design was based on the view that looking at a single belief-practice linkage will not necessarily yield valid findings because of the influence of the many considerations faced by a teacher in his/her teaching. Thus, for this study, I selected a variety of methods to enable me both to approach the main research question from several different perspectives and to uncover some of the other influences and interactions not represented in the initial conceptual framework. The ultimate goal was to enhance the depth of my understanding of these interactive processes and to ensure validity of the findings. To achieve this goal, the present study employed a research design with a series of interactions through which it was the teachers *themselves* who did the articulating and, as much as possible, the connecting of their beliefs to their classroom actions.

Another major consideration in the selection of data collection strategies was that, while espoused beliefs can be elicited through direct interview, beliefs-in-use (which may be unarticulated) are only discoverable through observation and analysis. In designing the research, attention was paid to being sensitivity to the different types of beliefs, both in data collection and in the analysis plan, as described below.

4. DATA COLLECTION

Three stages of data collection were involved in this study. Table B-1 summarises the focus of each of the stages. The procedures and their justifications will be detailed in the following sections.

Table B-1. Main stages of data collection of the study

Stage	Focus
1	Identifying and selecting potential subjects for the study
2	Classroom observation and probing teachers' curricular decisions
3	Probing teachers' beliefs about science, teaching and learning, as well as other related beliefs

4.1 Stage 1 – Identifying and selecting the subjects

The key informants. In the TAS, every 15 schools are monitored by a TAS coordinator who is an experienced teacher appointed by the public examining body, HKEA. The responsibility of the TAS coordinator is to ensure that teachers under his/her charge are complying with the requirements of the TAS. To do this, the coordinator has to inspect the teachers' record of work including lab manuals, assessment checklists, the number and nature of practical work assessed and students' reports. Through such inspections, the TAS coordinators should be able to gain a rough idea of the teaching of practical work of the teachers concerned.

At the outset of this study, I invited seven TAS coordinators to recommend teachers whom they regarded as practising good teaching of practical work; "good" in the sense that the teacher was either trying out novel practices, showing signs of giving thought to how to teach practical work, keeping very systematic records of the relevant documentation, and so on.

That is to say, information about the teaching of practical work in schools was dependent on informants who had indirect contact with the teachers concerned. Thus, the teachers selected for this study represented a specific group of teachers who were explicitly identified as those doing "good" teaching of practical work. As a result, any findings and interpretations of the case stories of these teachers have to be made with this in mind.

Selection of the subjects. I then approached the teachers recommended by the TAS coordinators and chatted informally with each of them individually. During the conversation, I tried to obtain information from the teachers about their teaching experience, the background of the schools in which they were teaching, the class size of their TAS classes and the nature of their students, and the characteristic features of their TAS practices. At the end of the conversation, I asked the teachers whether they would be willing to participate in the study if selected. Teachers who expressed initial consent were then requested to provide me with access to their instructional materials for teaching practical work. Subsequently, the teachers were invited to participate in the main study if they showed apparent willingness to share information with me on their own teaching and personal thinking on the teaching of practical work. This selection criterion was important in determining whether further involvement of the teachers in the study would enable me to examine their personal beliefs.

Of the teachers who seemed to satisfy the above criterion, I then tried to include in the final sample teachers whose practices seemed to show marked variations with each other. The idea was to include as much variation in teachers' practices as possible in the sample although there was no intention to claim that any of the findings would be generalizable to other settings. Differences in teaching experience and school background were also considered.

4.2 Stage 2 – Classroom Observations and Post-lesson Interviews

A total of ten teachers were invited to continue participation in this study. This was deemed to be about the right number when the amount of time and effort required for each case study and the amount of information needed for serious inferences to be made were considered. Relevant documents including the year plan, lab manuals, assessment checklists collected from each of the teachers were analyzed separately to draw out main themes and patterns. The idea was to get a general understanding of what the teachers' emphases were with respect to their teaching of practical work.

I then requested the teachers to provide me with observational access to their classrooms. For each teacher, lessons associated with at least four different practicals were observed, attempts were made to observe both TAS and non-TAS practicals. Not only the laboratory sessions were observed; other lessons associated with the practicals concerned were also observed, whenever possible. These included the pre- and post-laboratory sessions if they were implemented separately from the laboratory session proper.

During classroom observations, all verbal transactions between the teacher and the students were audiotaped using a remote wireless microphone worn by the teacher and a tape recorder controlled by me. I sat in at the back of the class in a corner of the classroom, monitoring the recording as well as jotting notes of the events taking place. Such an unobtrusive arrangement appeared to quickly desensitize both the teacher and the students to my presence as an outsider, and allowed the class to proceed as normally as possible. Field notes were made of different aspects of the lessons that were considered either to be representative of interactions or that raised questions to me about the teacher's practice, concerns or beliefs. These included, for example, the teacher's movements and apparent enthusiasm, general classroom tone, student behaviour and my impressions of the overall class proceedings. To facilitate the integration of these field notes with transcripts from the audiotapes, I frequently jotted the time and "counter number" from the tape recorder on the field notes. Supporting documentation such as lab sheets, assessment checklists and other supplementary materials used in the lesson were collected from the teacher before the start of the lesson. The field notes were used as a source of questions for the post-lesson interviews, descriptors of classroom practices, and of items for the Dilemma Interview and Decision Interview (see Appendix B 4.3).

The Post-lesson Interview (PIN). A post-lesson interview with each teacher after the observed lesson was carried out in order to clarify any specific observations related to that particular lesson. The post-lesson interviews followed a semi-structured format with an opening question, "Are there any special instances in this lesson that you want to discuss with me? " Each of the instances raised by the teacher (and, at a later point, those raised by me) was followed up by the following questions, wherever appropriate:

- What were you doing in this instance and why?
- What did you notice about the students?
- Did you have any particular objectives in mind in this instance? If so, what were they?

As soon as possible after each classroom observation, audiotapes of the lesson were listened to carefully as soon as possible to see if there were further instances that had escaped my attention during the classroom observation, and hence had not been asked of the teacher to justify or explain in the post-lesson interview. If this occurred, I interviewed the teacher about these further instances over the phone as soon as possible, in order that the teachers would still have a fresh memory of what they had done in their lessons. The totality of the field notes and the PIN transcripts became the data file of the lesson concerned.

My presence in the classrooms, and the fact that I made audio-recordings of the classroom events, provided an important signal to teachers that I was well-informed about the patterns of events taking place during the lessons. The implication here was that I was seen by the teachers as a well-informed observer who was not solely dependent on the teachers for information. This helped to minimize the possibilities open to the teachers for inventing plausible or misleading answers in the PIN. In sum, the purpose of the PINs was to allow the teacher to reconstruct thought processes in the context of the situation in which they actually occurred. The meanings and explanations teachers attributed to their experiences were crucial for understanding their classroom actions, and it was in the search for such meanings that the PIN became a valuable tool. I also tried to get a sense of whether the teacher's answers were representative of their underlying beliefs, or contained a problem or situation she/he wanted to address.

4.3 Stage 3 – Probing further into the teacher's beliefs

On completion of *all* the classroom observations for a particular teacher, the teacher concerned was interviewed on four occasions, and each interview had a different focus as indicated below:

- Dilemma Interview (DELM Interview) – based on 'dilemmas' experienced by the teacher as revealed in their responses to the questionnaire "Teaching of Practical Work at the Sixth Form Level"
- Decision Interview (DCSN Interview) – a semi-structured interview probing teacher's considerations when they are planning their practical work teaching
- Conception of Teaching Science Interview (CTS Interview) – a semi-structured interview probing teacher's conceptions of science, science teaching and learning, etc.
- Biography Interview (BIO Interview) – a semi-structured interview probing teacher's experiences both as a learner as well as a teacher

Timing of the interviews. There were three reasons for conducting the above interviews only after *all* classroom observations for the teacher concerned were over. Firstly, I was concerned that if these interviews (which were designed to probe the teachers' beliefs) were conducted before classroom observations, the teachers, having expressed their beliefs, might feel obliged to teach in ways that were congruent with what they had espoused earlier in the interviews. This might constitute a potential source of contamination for the classroom observation data. In contrast, I thought, if these interviews were to be held only after all classroom observations had been completed, this would have the advantage that teachers would be more likely to respond to interviewing questions based on actual examples and instances of his/her teaching during which I was also present. Thus, this shared experience with the teacher (by my having been in some of the lessons) provided a bridge or common language between us. This, in turn, would enable me to enter into the 'world' of the teacher for any interpretations I sought to make. In addition, through such familiarity, and that gained through the PINs, I hoped that teachers would feel sufficiently at ease to tell me their genuine beliefs rather than their espoused beliefs.

Second, if there were cases where the teacher's expressed beliefs were different from or conflicted with the observed classroom actions, it would then allow me to probe further into the discrepancies. In so doing, there would be a greater chance for me to reveal the teacher's genuine beliefs or obstacles that had prevented the teacher from acting according to his/her ideal. This was particularly true for the Dilemma Interview detailed below.

Third, in order to probe the teacher's beliefs about (1) the role of practical work and (2) their concerns in the teaching of practical work and its assessment, a Dilemma Interview and a Decision Interview were to be conducted with each of the teachers. For the same reasons mentioned above, namely, to enable me to enter into the 'world' of the teacher, a list of events of practical work teaching had to be derived from the actual classroom actions of all the teachers. That means, the list of classroom events necessary for composing the questionnaire (which would then serve as the interview protocol for the Dilemma Interview, see below for details) could not be made until the classroom observation stage was over. In short, there were both methodological concerns and logistic reasons for scheduling the various interviews after all classroom observations were completed.

The Dilemma Interview (DELM). It is not surprising that tensions occur within a teacher's beliefs about teaching science. Because of the complexity of classrooms, even if a teacher makes a major attempt to reconcile the many different aspects of teaching science, there are always likely to be competing demands that cannot be ignored (Lyons, 1990). It is in this context that the teacher is aptly portrayed as a dilemma manager, a broker of contradictory interests, who "builds a working identity that is constructively ambiguous" (Lampert, 1985, p.190) in order to combat these pedagogical dilemmas. Articulation of dilemmas often yields rich information regarding teachers' beliefs. This underpins the value of the Dilemma Interview in probing the teachers' beliefs behind their classroom actions, in particular, (1) their beliefs about the role of practical work in school science education, and (2) their concerns about the teaching of practical work and its assessment.

Several days before these interviews, each teacher was supplied with a questionnaire consisting of a list of events of practical work teaching (see Table B-2 for examples of the events) and asked to rate each of the events on two separate 5-point Likert scales according to: (1) the *importance* of such an event in the teaching of practical work in the sixth-form biology curriculum; and (2) the *frequency of occurrence* of such an event in his/her *own* teaching situation. At the end of the supplied list of events, the teacher was also asked to describe any additional event(s) which they thought was/were important and/or occurring frequently in their classroom. As mentioned above, the supplied list of events of teaching practical work was developed primarily based on all my observations of the teachers' lessons. The intention was to develop articulations of beliefs arising out of practice – that is, to access, as Argyris and Schon (1974) put it, "in-use" beliefs.

Throughout the interview, I stressed to the teacher that the focus of the interview was on his/her own teaching situation rather than on other peoples' situations. Based on the teacher's ratings of the importance and the frequency of occurrence of items in the list of events, the teacher was interviewed on the justification, reasoning and explanation for their ratings. In cases where there were discrepancies between the teacher's rating and what I had observed during the classroom observations, clarification was sought from the teacher and the rating amended, if deemed necessary. Special attention was paid to events where there were noticeable tensions (e.g. the teacher thinks that the event is important but he/she carries it out only very rarely).

Table B-2. Examples of events of practical work teaching extracted from the questionnaire for the Dilemma Interview

	Events of practical work teaching
1	Have a detailed year plan of practical work teaching set out at the beginning of the term
2	Conduct special sessions to introduce pupils to concepts like hypothesis, control, variable, etc. before engaging them in the actual practical work sessions
3	Distribute the lab manual to pupils in advance of the practical session
4	Inform pupils in advance whether a specific practical will be assessed or not
6	Have an in-depth pre-lab discussion with pupils (usually lasting for more than 15 mins)
7	Draw pupils' attention to lab safety as part of a laboratory session
16	Provide changes for pupils to investigate their own problems through practical work
18	Allow pupils to work in groups (say, in groups of 2 to 3) during the practical
19	Allow pupils to have discussion among themselves during the practical
21	Refer a pupil to consult his/her classmate with whom the teacher has previously discussed a similar problem
22	Refer a pupil to consult his/her classmate who is performing the task very well
23	Hint to a pupil that marks might be deducted from him/her if he/she were to proceed according to his/her planned course of action
24	Hint to a pupil that marks have already been deducted from him/her for committing a certain mistake
25	Question a pupil about the rationale behind a certain experimental procedure
26	Allow pupils to look up references during the practical
27	Initiate discussion with a pupil on the results he/she obtained during the practical
28	Avoid giving assistance to pupils during the practical
29	Avoid discussing with pupils during the practical
32	Prepare a detailed checklist for assessment purposes
37	Require pupils to observe a stringent deadline for handing in their reports
40	Conduct a post-lab discussion before pupils are asked to write up their own reports
41	Discuss with pupils the 'failure' of an experiment about which even the teacher himself/herself is uncertain of the possible causes
42	Make reference to the 'scientific method' when discussing with pupils
44	Make reference to the scientific attitude when discussing with pupils
45	Insist that pupils should report the actual data they obtained even if they are anomalous
51	Provide pupils with a handout highlighting the common mistakes found in their reports
52	Disclose the assessment criteria to pupils e.g. explain to them the marks they got for a certain assessment

The Decision Interview (DCSN). Like the Dilemma Interview, the Decision Interview was intended to develop articulations of beliefs arising out of the teacher's own practice so as to access their "in-use" beliefs. In its original design, teachers were asked to elaborate on their rationale for choosing to implement a particular practical activity in their teaching, so as to probe their beliefs about the role of practical work in school science teaching. However, results of a pilot test with other teachers revealed that some teachers found it difficult to articulate the rationale behind their curricular decisions. Instead, they found it much easier when provided with alternatives of what other teachers were doing for the same topic. Thus, I decided to postpone this interview, once again, until all classroom observations were over. This enabled me to compile, for each of the topics (seven in total), the whole spectrum of practical work that had been carried out by all teachers in the study.

One week prior to the Decision Interview, the teachers concerned were informed that, among other interviewing questions, they would be asked to give a narrative account in each of the following areas:

- A recent example of good quality practical work in my sixth form biology lessons
- A recent example of good quality practical work assessment in my sixth form biology lessons

In doing so, they should take into account of the whole teaching process; that is, including planning, organization, teaching approach and, if appropriate, assessment.

Teachers were informed of this requirement one week in advance so as to allow them sufficient time to recall and to reflect on their practices, rather than giving a spontaneous answer on the spot. An oral account was preferred to a written account as originally intended, mainly because of the concern not to overload teachers with too many demands from me, the researcher.

These narrative accounts were designed as an indirect means of probing the teacher's beliefs about the role of practical work and the role of practical work assessment. It was assumed that these accounts would generate indications of the key elements in their knowledge and beliefs about the role of practical work and its assessment. This indirect way of probing the teachers' beliefs was thought to be superior to asking them directly about their views of the respective areas. In the latter case, the teachers would be more likely to respond by giving out 'standard' answers that they had learnt in education courses. Or, they would simply regurgitate some of the grand aims that are cited in the official curriculum guides. In the former case, on the other hand, the teachers have to illustrate their views with specific examples. Through the presentation of reasons why the examples are of good quality practical work, and through reflection on the relationship between means and ends, the knowledge claimed by the teacher becomes more credible and valid. Fenstermacher (1994) calls this reflection on practice, practical reasoning, which he defines as a means for transforming the tacit quality of the teacher's knowing to a level of awareness that opens up the possibility for reflective consideration. This, he argues, is the minimal form of warrant for practical action.

Notwithstanding the measures taken, this indirect way of probing the teacher's belief still cannot guarantee validity of the data collected. Along with the proliferation of narrative studies have been constant expressions of concern about the validity and meaning of the stories collected. Leinhardt (1990) contends that it is hard to know whether the teacher has concocted some "magical brew" to describe what he or she is doing, or to determine whether the teacher is reporting the critical pieces of performance. Phillips (1993), in decrying the lack of epistemological considerations in narratives, makes a similar point – "the fact that a story is credible tells us nothing – absolutely nothing – about whether it is true or false" (p.21). Thus, the utilisation of narrative accounts as a valid method for the explication of teacher beliefs is only acceptable to the extent that the beliefs provided are, in some way, warranted as described above. In the present study, the warrant was further provided by asking the teachers to reflect on their personal life experiences in biology teaching in general through the Biography Interview, details of which will be described below.

In summary, the narrative accounts of good quality practical work and assessments were themselves indirect ways to probe the teachers' beliefs concerning their corresponding roles in biology teaching. Probing through these indirect means minimized the chances of teachers reproducing 'standard answers' that they might have come across in curriculum guides or the TAS Handbook. The validity and trustworthiness of the narrative data were further increased by triangulating them with other sources of data, that is, those from the other kinds of interviews such as the Biography Interview.

During the interview, the teacher was first asked to give narrative accounts of a recent example of good quality practical work and good quality practical work assessment. At appropriate points, I chipped in to clarify and probe further into the teacher's beliefs about the role of practical work and its assessment in science teaching. Then the teacher was shown, for each of the topics (seven in total), a list of experiments and practical activities that had been provided to students, either by the teacher himself / herself or by other teachers in the study. The teacher was asked to comment on the rationales behind his/her particular curricular choices with respect to (1) why a particular experiment was chosen, (2) the way in which the practical work had been implemented, and (3) why it was assessed or why it was not assessed. In short, the focus of the interview was on probing the teacher's beliefs about the role of practical work and its assessment in science teaching. Table B-3 shows, as an illustration, the list of experiments that was used to elicit teachers' curricular choices with respect to their teaching of one of the topics – enzyme reaction.

Table B-3. List of experiments used to elicit teachers' curricular choices in teaching the topic enzyme reaction

	Title of the experiment
1	Compare the protease activity of papaya, pineapple and kiwi fruit (using milk agar as the substrate)
2	Compare the protease activity of papaya, pineapple, orange, apple, biological washing powder and meat tenderizer (using milk agar as the substrate)
3	Compare the protease activity of papaya, pineapple, orange, apple, biological washing powder and meat tenderizer (using photographic negative as the substrate)
4	To show that the digestive effect of kiwi fruit extract is concentration dependent (using milk agar as the substrate)
5	To demonstrate the digestive effect of trypsin (a protease), kiwi fruit extract and biological washing powder on milk agar
6	To show that the digestive effect of biological washing power is due to an enzyme (using milk agar as the substrate)

The Conception of Teaching Science (CTS) Interview. The CTS interview, originally developed by Hewson and Hewson (1989), was chosen as a tool that would provide detailed analyses and descriptions of the teachers' conceptions of teaching science. The notion of a conception of teaching science recognizes that the knowledge used in teaching draws from knowledge of the subject, the learner, and instruction, and it is sensitive to the particular context in which a teacher uses it. The interview consisted of ten events that included instances and non-instances of science teaching and learning both in and out of classroom contexts, dependent upon teachers' interpretations. The description of the ten events had been modified slightly and pilot tested so that they fitted better into the local context (see Table B-4). In the interview, each teacher was shown a written description of each event in the same sequence. For each event, the teacher was asked the following questions:

- In your view, is there science teaching happening here?
- If you cannot tell, what else would you need to know in order to be able to tell? Please give reasons for your answer.
- If you answered "yes" or "no", what tells you that this is the case? Please give reasons for your answer.

I introduced no other terms into the interview, but did follow up on ideas (for example, about learning) introduced by the teacher. The interview format allowed the teacher to be more reflective in expressing their ideas, allowing opportunities for reconsideration of earlier statements in light of later instances. Ample opportunities were given to the teacher to provide reasons for their judgment. In short, the CTS Interview task is designed to enable respondents to consider the components of an appropriate conception of teaching science (Hewson and Hewson, 1988), at the same time providing an environment in which a variety of views can be expressed without bias from the task structure. The task does not prescribe what is important, nor to which aspects attention should be given. A number of instances are, in fact, ambiguous; in other words, they require respondents to be able to contribute their own ideas and focus on what is important or significant to them. It is developed to allow subjects to respond to particular events, while encouraging them to link the events to larger conceptual issues (Hewson et al., 1995). It was my intention to use this interview to generate additional, confirming or qualifying, information about each teacher's views of science, science teaching and learning.

Table B-4. List of events for the Conception of Teaching Science Interview Task

Event	Description
1	Teacher in Form Six at the start of the topic on arthropods, passes around a box of specimens, containing insects and spiders, teachers asks, "What can you tell about these specimens?"
2	A student at home watching a TV programme (Life on Earth) on how different organisms adapt to different ways of life.
3	Two students working together in the library doing calculations on problems concerning calorific values for different foods.
4	A professor lecturing on Darwin's theory of natural selection to a group of primary one students.
5	Teacher in front of a Form Six Biology class describing the steps in using the Punnett square method of figuring genetic offspring ratios.
6	Teacher reads a biology student's statement that tomatoes are vegetables and asks, "Is there any difference between biological definitions of fruits and everyday definitions of fruits and vegetables?"
7	Teacher at the end of a demonstration of skeleton distributes a drawing and asks students to label the main bones from memory.
8	Junior secondary school student in class looking at a chart showing arterial blood as red and venous blood as blue, asking the teacher, "How does the blood change colour?"
9	A student at home following a recipe for making bread using yeast and flour.
10	Teacher writing a self study resource centre programme on how to use the microscope to estimate the size of a stomatal pore.

The Biography Interview (BIO). This was the last interview for each teacher. It provided an opportunity to explore the teacher's background, focusing on their general feelings about themselves as students and as teachers, their general reactions to their own science backgrounds, their roles as teachers and, more specifically as science teachers, their beliefs about how children learn and their ideals of science teaching. It was a semi-structured interview guided by a number of questions (see Table B-5). The design of this Biography Interview was predicated on the stance of the present study to adopt a holistic approach to the phenomenon under study and to take into consideration the goals pursued by the individual teachers. This necessitated probing the teacher's biography or personal experiences that may have influenced his/her classroom action.

Table B-5. Examples of interview questions for the Biography Interview protocol

	Questions
1	Describe for me one of the best learning experiences you had as a student.
2	Describe for me one of the worst learning experiences you had as a student.
3	Do you take courses now which you enjoy or which you feel you should enjoy? Give me one or two examples of each kind?
4	Did you hold any other jobs before teaching? Describe.
5	Describe what it is about teaching that you really like.
6	Describe what it is about teaching that you really don't like.
7	How would you characterize yourself as a science teacher?
8	Describe for me the ideal teacher of Secondary Six level teaching biology.
9	Have you changed your ideas about what constitutes good science teaching over the years? If so, how? What do you think caused the change?
10	Describe for me the ideal student for science at Secondary Six level?
11	Have you changed your ideas about how students learnt best over the years? If so, how? What do you think caused the change?
12	Are there any constraints, or external factors within the school, beyond your control, which limit the way you would like to be teaching biology?
13	What are some examples of ways in which you feel you do a good job in science teaching?
14	Has the TAS had any effect on your ideas about teaching and your actual teaching since its implementation?

4.4 Summary of data collection methods

Figure B-1 summarizes the various data collection methods of this study and indicates how each of the methods contributed to the six data pools required for answering the research questions. The arrows represent the major contributions of each data collection method to the various data pools. In reality, the five different kinds of interviews (i.e. PINs, DELM, DCSN, CTS, BIO), because of their semi-structured nature, often provided data for other data pools as well. For the sake of presentational clarity, for each of the interviews only the major contribution(s) is/are shown.

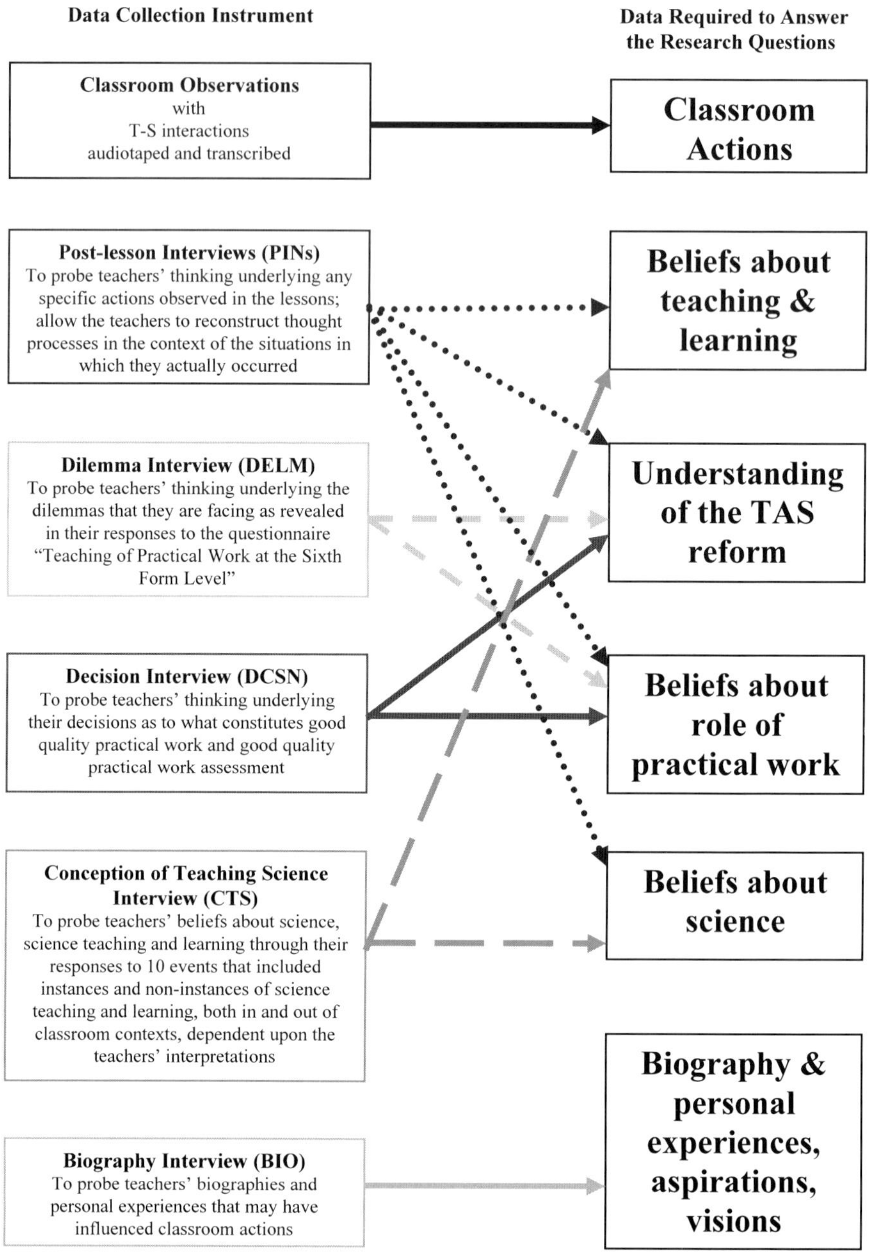

Key: the arrows represent the major contributions

Figure B-1. Overview of the data collection instruments

5. TREATMENT AND ANALYSIS OF DATA

Altogether, data from six main sources were collected during the field study, each of which was also supplemented by relevant field notes and/or teaching materials such as lab manuals and assessment checklists. In chronological order of their collection, they were: (1) audiotapes of dialogues between the teachers and their students during the practical lessons, (2) the Post-lesson Interviews, (3) the Dilemma Interviews, (4) the Decision Interviews, (5) the Conception of Teaching Science Interviews and (6) the Biography Interviews. Overall, a total of 52 classroom observations of practical lessons were conducted, each lasting between about two to four hours; and a total of 92 interviews were conducted, each lasting between half an hour to two hours. All the above were transcribed and translated verbatim before subjecting them to further analysis. This amounted to more than 1000 single spaced pages of transcripts of classroom data and interview data, all of which were entered into the computer software QSR NUD*IST (1997) to facilitate subsequent analysis.

The data collected were, in the main, analysed qualitatively although some simple quantitative analyses on the classroom data (as reported in Chapter 3) were also carried out. Presented below are the main stages and broad approach to data analysis determined prior to the study. As with most interpretive or qualitative research, many of the details of the coding systems were refined at a later stage after the field study had been completed and the data examined for trends and patterns.

5.1 Interview data

Analysis of the interview data preceded that of the classroom data. This was thought to be appropriate as a deeper understanding of the teachers' beliefs about science, and those of teaching and learning as a result of the analysis of the interview data, would shed light on how best to analyze the classroom data. Hence, there would be a greater chance of teasing out the relationships between the teachers' classroom actions and their underlying beliefs.

Altogether, three rounds of analysis of interview data were undertaken. Each represented an attempt to search for the answer to a particular research question. The ***first round of analysis*** was conducted in three phases. At the beginning, the examination of the transcripts was mainly guided by the initial interpretive framework described in section 2 above, which was designed for abstracting data to answer research question 3: What are the teachers' beliefs about science, teaching and learning? Within this interpretive framework, categories were generated by an examination of the data and attempts to identify common themes. This was the constructive phase of data analysis during which examination of the transcripts facilitated creation of initial categories. The second phase involved integration of these categories and their properties. During this phase, similarities and differences among the categories created in phase one were identified. Some categories were combined with others that had similar properties. In the third phase, the data were further integrated to form fewer, more encompassing categories.

This whole of the first round entailed creating new categories, refining and sharpening categories, and elaborating on existing categories. The three phases did not follow in a simple linear progression. Typical of this method of analysis, the phases formed a repetitive process of coding, comparing, and refining (Glaser and Strauss, 1967). This first round of data analysis contributed to preparation of the portraits of individual teachers reported in Chapters 4 to 11, and the combined summary for all the ten teachers in Table 12-1.

The ***second round of data analysis***, using a similar procedure, was repeated to abstract data from the transcripts to answer research question 2: What understanding / perceptions do teachers have of the TAS? However, this round of data analysis differed from the first one in that none of the constructs in the initial interpretive framework guided the analysis. Instead, the interpretive framework was arrived at as a result of several attempts at data examination, reduction and display. The resulting interpretive framework comprised three constructs, namely:

- The TAS as an assessment reform
- The TAS as a pedagogical reform
- The TAS as effecting teachers' professional development and professionalism

In the ***third round of data analysis***, again through similar procedures, two new constructs – teachers' views of fairness and their sense of professionalism – were developed, enabling the researcher to probe further into the data in an attempt to answer research question 4: What are the relationships among the teachers' beliefs about science, teaching and learning, their understanding of the TAS reform, and the ways they implement the reform inside their classroom?

At this juncture, readers need to be reminded that although each of the interviews (PINs, DELM, DCSN, CTS, BIO) had their own foci and emphases, there were also overlaps because teaching of practical work does not occur in a vacuum. For example, when a teacher mentioned something about the role of practical work in the Decision Interview, he/she might also link it with belief about learning or about the nature of science. Therefore, analyzing the data collected from all the *different* sources by using the *same* set of interpretative frameworks (one at each of the three rounds of data analysis) provided a convenient way to integrate and synthesize the disjointed pieces of information into an integrated whole. In the process of integrating the data, I was also triangulating the data collected from the different sources.

A point of clarification needs to be made here regarding the different types of interview data. Data from the Post-lesson Interviews (PINs) differed from the rest in its adoption of a naturalistic approach. The focus of PINs was on how teachers made sense of their own classroom actions through stimulated recall by the researcher. This was distinctly different from the rest of the interviews in which teachers were making sense of their actions in the TAS *indirectly* through their responses to interview questions based on:

- their responses to a questionnaire on their classroom actions (DELM)
- interviewing tasks probing their decisions in teaching practical work (DCSN)
- interviewing tasks probing their conceptions of teaching science(CTS), and
- semi-structured interview questions on their biography (BIO)

In the context of the current study, these two approaches complemented each other. On the one hand, the naturalistic approach, with its emphasis on studying direct experiences taken at face value, has the advantage of examining the reality through the teachers' eyes and hence the possibility of producing authentic and trustworthy data. However, it relies on the sensitivity (and skill) of the researcher in raising questions with the subjects in order that relevant responses can be elicited. On the other hand, during the process of data collection it was noted that the data collected by the latter group of interviews was able to reveal some of the teachers' thinking that would otherwise have remained undetected by simply relying on the PIN Interviews. In particular, the data revealed that the DELM Interviews were especially useful in uncovering some of the teachers' thinking behind classroom actions that they had tried to avoid. Indeed, most of these 'avoidance behaviours' did not readily manifest itself outwardly in terms of observable teachers' behaviour. As such, this made it very difficult, if not impossible, for the researcher to notice, and to raise the related issues with the teacher during the subsequent PIN Interviews. Examples of these kinds of classroom actions included the following:

- Avoiding distributing lab manual in advance of the practical session
- Avoiding initiating discussion with individual students during the practical work session
- Avoiding giving assistance to individual students during the practical work session

5.2 Classroom data

After all the interview data were analysed, and with the benefit of having a basic understanding of the various beliefs held by the teachers, the transcripts of the classroom data of each of the teachers were then studied over and over again. The aim was to develop a coding scheme which would enable the abstraction of the classroom data in such a form that could answer research questions 1 and 4. That is: What are the characteristic features of secondary science teachers' classroom practice in the TAS? And what are the relationships among the teachers' beliefs and their classroom actions?

All the classroom data were coded with the help of the QSR NUD*IST programme (1997). As reported in Chapter 3, essentially, each DTU has at least two codes – one for the *focus* and the other for the *type* of T-S interaction. In order to check for possible coder bias, an experienced biology teacher was recruited as a 'judge' to code the transcripts of classroom data of five randomly selected practicals. The inter-coder agreement between myself and the experienced teacher for the coding of *foci* and *types* of T-S interaction was 87% and 91% respectively. These were deemed to be satisfactory, given the ways in which the data were then used. Further details about the method of analysis are given in Chapter 3, together with findings concerning the characteristic features of the teachers' classroom practice in the TAS, and will not be repeated here.

6. REPORTING THE DATA AND THE FINDINGS

6.1 Classroom actions

In order to study the relationship between teachers' classroom actions and their underlying beliefs productively, two questions need to be addressed. First, how might teachers' classroom actions be characterized? Second, what evidence of teachers' classroom action is the most critical for subsequent analysis? These questions were addressed by providing readers with an overall picture of the classroom actions of the ten teachers in this study in a quantitative manner, in Chapter 3, and highlighting those features that were most critical for subsequent analysis. Specific examples of how the classroom actions of individual teachers were related to their underlying beliefs were left to subsequent chapters, where they are reported in a more qualitative manner in the form of case stories for individual teachers. In the main, the report in Chapter 3 addresses research question 1. That is, what are the characteristic features of secondary science teachers' classroom actions in the TAS?

6.2 Case stories for each of the teachers

The case stories in Chapters 4 to 11 report on the possible relationship between the teachers' classroom practice and their underlying beliefs. They also focus on reporting the teachers' perceptions / understanding of the TAS as revealed in the reasons they gave for their classroom actions. Thus, they address all the four research questions in the form of a case report for each of the teachers. In order to help readers to develop a holistic picture of the relationship between the teachers' classroom practice and their beliefs, each case story begins with a portrait of the teacher, comprising his/her personal, educational and professional backgrounds, his/her attitudes and aspirations, as well as his/her beliefs about the nature of science, the role of practical work, and those of teaching and learning. These descriptions mainly contain information collected from the BIO, CTS, DELM and DCSN Interviews. In other words, the beginning parts of the case stories address research question 2, namely: What are secondary science teachers' beliefs about science, teaching and learning?

6.3 Teachers' different views of fairness

By reporting how three cases are tied together by the teachers' concerns with the notion of fairness, the relationship between the teachers' classroom actions and their underlying beliefs is explicated in Chapter 13. In a way, this chapter addresses all four research questions at the same time, and in a holistic manner, by comparing the differences among the three cases.

6.4 Teacher professionalism and policy interpretation

By reporting how five cases are tied together via the teachers' interpretations (albeit different) of the same policy statements, Chapter 14 attempts to tease out the relationships among the teachers' sense of professionalism, their interpretation of the reform policy and the ways they implement the policy inside their classroom. Again, this chapter addresses all four research questions at the same time, and in a holistic manner, by comparing the differences in the cases.

6.5 Selection of cases for reporting

In the course of selecting the cases for reporting, tension arose in choosing which cases to be reported, and in what detail. Such tension arose from the nature of the case study methodology, which led to thick descriptions based on the classroom observations and teacher interviews. Limitation of space only allowed the reporting of eight of the cases with any degree of detail. These case reports serve as exemplary cases to illustrate the theoretical frameworks concerned, the analytic method and techniques used in this research, and also to demonstrate the evidence that warrants the inferences and assertions made from the data collected.

It should, however, be pointed out that the data reported in the selected case stories differ in their richness and the relative amount of information on the themes discussed. Such differences are the result of a number of factors, including the number of lessons observed, and the amount of information offered by the teacher during the interviews. The latter depended in part on the willingness of the teacher to talk, how expressive or articulate the teacher was, and how much the teacher had to offer. Although I had intended to obtain information on various aspects of the teacher's beliefs, the flow of interviews was at times influenced by what the teacher chose to talk about.

7. REFLECTIONS ON METHODOLOGICAL ISSUES

7.1 Approach to validity

A major issue in any research project concerns the validity or trustworthiness of the data and its interpretation. In this study, triangulation was the strategy used to enhance the validity of assertions, which involved the use of divergent means for gathering evidence across a range of different kinds of *sources* and *techniques*. Data were obtained from *three sources*: classroom events, curriculum materials used by the teachers and interviews with teachers. *A variety of techniques* were employed – data gathering from documents, classroom observation, recording of teacher-student interaction, interviews based on stimulated recall of classroom events (Post-lesson Interviews), an interview based on standardized interviewing tasks (Conception of Teaching Science Interview), an interview based on teachers' responses to a questionnaire (Dilemma Interview), and a biography interview, etc.

All sources of data, and the results of data analysis from each phase, were considered in totality to help judge the congruence of the overall case study developed for each teacher and to answer the research questions. For example, general statements concerning the role of practical work in science education made by a teacher in the Decision Interview were judged against what actually occurred in the classroom situation and against similar statements described in various other interviews. Thus, the use of numerous data sources (collected using different techniques) maximised the probability that the emergent assertions were robust. In order to ensure the credibility of the findings, any evidence that was counter to an assertion was explored in detail, and care was taken to explain the discrepancy. In this process, assertions were elaborated so as to take account of all data.

Every assertion developed from data was analyzed critically, and the data was probed, and, in some cases, the teacher concerned, for other possible explanations and counter-examples. The various interviews, in particular, the Post-lesson Interviews, the Decision Interviews and the Dilemma Interviews were especially useful because they encouraged the teachers to do much of the sense-making and the looking for meanings; my role as the researcher was to get the teacher concerned to clarify and interpret where possible.

The extent of transferability of the findings (i.e. external validity) of this study to contexts that readers might encounter will be constrained by the perspective of the reader and the potential applications the reader might consider. Readers have to decide for themselves what are the potential applications, if any, of the findings to their own context. For this reason, it is important to delimit the extent to which findings from this study can be applied through a "thick description" for each of the case studies, so as to provide sufficient information for readers to assess the potential transferability and appropriateness for their own settings (Guba and Lincoln, 1989). To this end, narrative accounts, in the form of case stories, were used to convey the context of this study and the knowledge that was implicit in the stories of the teachers. In this way, I think, the voices of the teachers can be heard in a more authentic and credible manner.

In all, I believe that, in their totality, the case stories do demonstrate the use of the conceptual tools and constructs and the process of analysis. They also present the range of evidence upon which my interpretations and analyses were made. As such they should enable readers of this book "to function as a co-analyst" (Erickson, 1986, p.145-6) of the cases reported, to understand the cases and to judge the validity of my interpretive analysis.

7.2 The Dilemma Interviews

I was particularly pleased with the Dilemma Interviews, especially in the way that they were conducted. The interview had been modified so as to be grounded in actual observed classroom practices as far as possible. In so doing, these interviews were more powerful than standardized interview protocol because they relied on classroom observations. Nevertheless, they had the drawback that they were largely dependent on my ability to capture salient classroom events that would also elicit significant responses from the teachers.

In many cases during the interviews, the teachers confirmed that they recognized their classroom behaviours when they looked at the questionnaire items. They also naturally referred to teaching events in lessons during which I was present to further illustrate or explain their points. Thus, the teacher and the researcher entered into a mutual understanding, with the observed lessons forming a reference point, or bridge for exchanging their perspectives. This was important because this mutual understanding did enable me to get a more accurate interpretation of the teachers' teaching and their underlying beliefs. Equally important, in order for me to have effective collaboration with them, was the teachers' need to feel that they were understood and their actions were to be interpreted correctly. In fact, it was those items of interview data which were most thoroughly grounded in classroom events, and expounded with consistency and intricate details, that comprised the most useful and interesting data.

7.3 An insider perspective and members check

I was Subject Officer (Biology) of the HKEA when the HKAL Biology TAS was in its initial stages of development, and subsequently became the Supervisor of the scheme when I left HKEA to pursue an academic career in a university. I resigned from the post of Supervisor before the conducting of this research in order to ensure that the data collected for the study would not be distorted in any way by my vested interests in the TAS. In fact, this point was made very clear to the participants of the study, as was their anonymity.

There were both advantages and disadvantages for me as the researcher in having such close connections to the TAS at the different stages of its development. On the positive side, this enabled the examination of the problem from an insider perspective, particularly in terms of my familiarity with the history and philosophies behind the development of the TAS. Such an insider perspective is important, especially in qualitative studies of this kind. On the other hand, the drawback is that bias was more likely to occur. Member checks (Guba and Lincoln, 1989) were thus instituted as one procedure to guard against this. I believe that the teachers' agreement with their portraits and individual case stories, written almost a year after the field study, provided a measure of validity of these results. Nonetheless, I am aware of the fact that, as in any interpretive research, the rich qualitative data collected in this research are open to more than one interpretation, but every effort has been made to provide defensible arguments for my interpretation of the findings.

Overall, the research design appears to have been effective in eliciting beliefs and descriptions of practices representative of the ten teachers. While no researcher can ever be absolutely sure of accessing accurately any teacher's thinking, the teachers in this study seemed motivated to give honest responses. The detail, precision and logical consistency of the teachers' explanation of their own work are impressive, and strongly suggest authenticity. Ivor's reference on several occasions to his dishonesty in administering the TAS was a case in point. In this respect, I am indeed most grateful to the teachers. Another gratifying thing for me in the course of this study was the willingness of these teachers to work with me as intensely as they did. The only real area of frustration was that the teachers sometimes articulated a desire for feedback and assistance (since I happened to be the ex-TAS Supervisor and a science teacher educator) which, of course, was not appropriate in the context of the research. The teachers understood this, but could not help occasionally asking for it.

7.4 The role of the researcher and that of the computer

In terms of the data analysis, as with the design, the approach taken appears to have been successful. The cross-case analyses reported in Chapters 13 and 14 were by far the most challenging. Without the help of the QSR NUD*IST programme, I would not have been able to easily retrieve, amidst the 'ocean of transcripts', the classroom episodes and interview transcripts that were related to the notion of fairness in one way or another. After further examination of these episodes and interview excerpts, I was able to tease out the different views of fairness that were associated with the teachers' classroom actions. The same applied to teasing out the relationship between the teachers' sense of professionalism and their interpretation of the TAS regulations. In all, the present study benefited considerably from processing the data using the QSR NUD*IST programme, though a great deal of time had to be invested to initially code the relevant excerpts. If this had not been the case, I cannot imagine what the situation would have been like. Probably, it would be something like rowing a boat in the middle of the ocean (of more than 1000 pages of transcripts). Nevertheless, I believe that the ultimate success of any research, in particular qualitative studies like the present one, lies with the conceptual apparatus that is used to abstract and interpret the data. In no way can the role of educational researcher be replaced by any super computer.

7.5 Implications for research

The current study has affirmed the value of a qualitative multiple-case study in its illuminative power to help one to understand the processes of change from the *teachers' own perspectives*. Findings reported in this book depend heavily on both classroom observational data and teacher explanations of their classroom actions. Much of the existing research on teacher thinking relies primarily on analyses of teacher statements about their practice unaccompanied by classroom observation. Without my close attention to the observational data in this study, I would have missed the importance of some of the teachers' avoidance behaviour and thus their different views of fairness and its impact on their classroom actions. Although the teachers' talk did come close to revealing these, their statements alone might have led to a simplification of their meanings. It was the observational data which provided strong support for the inferences that I was able to make.

The observations also allowed for the definition of *teacher* concepts. For example, all teachers talked about the notion of fairness, but the observational data indicated that they were referring to qualitatively very different ideas underlying the notion of fairness. In addition, the classroom observational data were found to be a very useful context for the subsequent interviews. The actual teaching episodes were used as concrete examples of the teachers' actions on which to base some of the questions for the interviews, and the teachers were asked to explain their actions. Starting from concrete and specific examples was an easier task than having teachers begin from a more abstract and philosophical statement of curriculum or teaching. All in all, one clear implication of this research is that *research on teacher thinking must involve observation of what teachers do in the classroom.*

References

Abd-El-Khalick, F., Bell, R.L. and Lederman, N.G. (1998). The nature of science and instructional practice: Making the unnatural natural. *Science Education, 82,* 417-436.

Airasian, P.W. (1988). Measurement-driven instruction: a closer look. *Educational Measurement: Issues and Practice, 7(4),* 6-11.

Andrew, S. (1994). The washback effect of examinations – Its impact upon curriculum innovation in English language teaching. *Curriculum Forum, 4(1),* 44-58.

Argyris, C. and Schon, D.A. (1974). *Theory in practice.* San Francisco: Jossey-Bass.

Armstrong, H.E. (1898). The heuristic method of teaching, or the art of making children discover things for themselves. In Board of Education, *Special Reports on Educational Subjects,* Vol. 2, pp.389-433. London: HMSO.

Atkinson, E.T. (1990). Learning scientific knowledge in the student laboratory. In E. Hegarty-Hazel (ed.), *The student laboratory and the science curriculum,* pp.119-131. London: Routledge.

Barthes, R. (1976). *The Pleasure of the Text* (trans. Richard Miller). London: Cape.

Bell, B. and Cowie, B. (2001). *Formative Assessment and Science Education.* Dordrecht: Kluwer.

Biggs, J.B. (1995). Assumptions underlying new approaches to educational assessment. *Curriculum Forum, 4(2),* 1-22.

Biggs, J.B. (1996). The assessment scene in Hong Kong. In J. Biggs (ed.), *Testing: To Educate or to Select? : Education in Hong Kong at the Crossroad,* pp.3-12. Hong Kong: Hong Kong Educational Publishing.

Black, P. (1998). *Testing: Friend or Foe?* London: Falmer Press.

Black, P. (2004). Personal communication with Peter Fensham. Cited in P. Fensham (2004). *Defining an Identity: The evolution of science education as a field of research,* p.173. Dordrecht: Kluwer.

Black, P. and Atkin, J.A. (1996). Changing the Subject: Innovations in Science, Mathematics and Technology Education. Routledge: London.

Black, P., Harrison, C., Lee, C., Marshall, B. and Wiliam, D. (2003). *Assessment for Learning: Putting It into Practice.* Berkshire: Open University Press.

Black, P. and Wiliam, D. (1998). Assessment and Classroom Learning. *Assessment in Education, 5(1),* 7-74.

Gilmore, A. (2002). Large-scale assessment and teachers' assessment capacity: Learning opportunities for teachers in the National Education Monitoring Project in New Zealand. *Assessment in Education, 9(3),* 343-361.

Gipps, C.V. (1986). *The GCSE: An Uncommon Examination.* University of London, Bedford Way Papers No. 29.

Gipps, C.V. (1994). *Beyond Testing: Towards a Theory of Educational Assessment.* London: Falmer Press.

Glaser, B.G. and Strauss, A.L. (1967). The Discovery of Grounded Theory: Strategies for Qualitative Research. Chicago: Aldine.

Goodson, I. (1997). 'Trendy theory' and teacher professionalism. In A. Hargreaves and R. Evans (eds), *Beyond Educational Reform: Bringing Teachers Back In,* pp.29-43. Buckingham: Open University Press.

Gott, R. and Duggan, S. (1994). *Investigative Work In The Science Curriculum.* Buckingham: Open University Press.

Grant, M. (1989). GCSE in Practice: Managing Assessment Innovation. Windsor: NFER-Nelson.

Guba, E. and Lincoln, Y.S. (1989). *Fourth Generation Evaluation.* Beverly Hills: Sage.

Gudmunsdottir, S. (1991). Ways of seeing are ways of knowing: The pedagogical content knowledge of an expert English teacher. *Journal of Curriculum Studies, 23(5)*, 409-421.

Hamilton, L.S. (2003). Assessment as a policy too. *Review of Research in Education, 27,* 25-68.

Hamilton, L.S. and Koretz, D.M. (2002). Tests and their use in test-based accountability systems. In L.S. Hamilton, B.M. Stecher, and S.P. Klein (eds), *Making sense of test-based accountability in education,* pp.13-49. Santa Monica, CA: RAND.

Hargreaves, A. (1989). *Curriculum and Assessment Reform.* Open University Press.

Hargreaves, A. (1992). Forward. In A. Hargreaves and M. Fullan (eds) *Understanding Teacher Development.* London: Cassell.

Hargreaves, D.H. (1994) The new professionalism: the synthesis of professional and institutional development. *Teaching and Teacher Education, 10(4),* 423-438.

Harland, J. (1988). Running up the down escalator. In D. Lawton and C. Chitty (eds), *The National Curriculum,* pp.87-98. London: Institute of Education.

Harlen, W. (1994). Issues and approaches to quality assurance and quality control in assessment. In W. Harlen (ed.), *Enhancing Quality in Assessment,* pp.11-25. London: BERA.

Harlen, W. and James, M. (1997). Assessment and learning: differences and relationships between formative and summative assessment. *Assessment in Education, 4(3),* 365-379.

Hawthorne, R.K. (1992). *Curriculum in the Making.* New York: Teachers College Press.

Hashweb, M.Z. (1996). Effects of science teachers' epistemological beliefs in teaching. *Journal of Research in Science Teaching, 33(1),* 47-63.

Helsby, G. (1995). Teachers' construction of professionalism in England in the 1990s. *Journal of Education for Teaching, 21(3),* 317-332.

Helsby, G. and Knight, P. (1997). Continuing professional development and the National Curriculum. In G. Helsby and G. McCulloch (eds), *Teachers and the National Curriculum,* pp.145-162. London: Cassell,

Hewson, P.W. and Hewson, M.G.A'B. (1988). An appropriate conception of teaching science: A view from studies of science learning. *Science Education, 72*, 597-614.

Hewson, P.W. and Hewson, M.GA'B. (1989). Analysis and Use of a Task for Identifying Conceptions of Teaching Science. *Journal of Education for Teaching, 15(3)*, 191-209.

Hewson, P.W., Kerby, H.W. and Cook, P.A. (1995). Determining the Conceptions of Teaching Science Held by Experienced High School Science Teachers. *Journal of Research in Science Teaching, 32(5),* 503-520.

Hillock, G. (1999). *Ways of Thinking, Ways of Teaching.* New York: Teachers College Press.

Hodson, D. (1988). A critical look at practical work in school science. *School Science Review, 70(256),* 33-40.

Hodson, D. (1992). Assessment of practical work: some considerations in philosophy of science. *Science and Education, 1,* 115-144.

Hodson, D. (1993). Against skills-based testing in science. *Curriculum Studies, 1(1),* 127-148.

Hodson, D. (1996). Practical work in school science: exploring some directions for change. *International Journal of Science Education. 18*(7), 755-760.

Hodson, D. (1998). Science fiction: the continuing misrepresentation of science in the school curriculum. *Curriculum Studies, 6(2),* 191-216.

Hodson, D. (2001). Research on practical wok in school and universities: In pursuits of better questions and better methods. Proceedings of the 6th European Conference on Research in Chemical Education, University of Aveiro, Aveiro, Portugal.

Hodson, D. and Bencze, L. (1998). Becoming critical about practical work: changing views and changing practice through action research. *International Journal of Science Education, 20(6),* 683-694.

Hofstein, A. and Lunetta, V. (1982). The role of the laboratory in science teaching: Neglected aspects of research. *Review of Educational Research, 52,* 201-217.

Hofstein, A. and Lunetta, V. (2004). The laboratory in science education: Foundations for the twenty-first century. *Science Education, 88,* 28-54.

Hong Kong Examinations Authority (1994). *The Work of the Hong Kong Examinations Authority – 1977-1993.* Hong Kong: Hong Kong Examinations Authority.

Hong Kong Examinations Authority (1996). Hong Kong Advanced Level Examination: regulations and syllabuses. Hong Kong: Hong Kong Examinations Authority.

Hong Kong Examinations Authority (1997). *Hong Kong Advanced Level Biology Teacher Assessment Scheme*. Hong Kong: Hong Kong Examinations Authority.

Hoyle, E. (1974). Professionality, professionalism and control in teaching. *London Educational Review, 3,* 13-19.

Hoyle, E. and John, P. (1995). *Professional Knowledge and Professional Practice.* London: Cassell.

Jones, A.T., Simon, S.A., Black, P.J., Fairbrother, R.W. and Watson, J.R. (1992). *Open Work in Science: development of investigations in schools.* Hatfield: Association for Science Education.

Kagan, D.M. (1990). Ways of evaluating teacher cognition: Inferences concerning the Goldilocks Principle. *Review of Educational Research, 60*, 419-469.

Kagan, D.M. (1992). Professional growth among preservice and beginning teachers. *Review of Educational Research, 62*, 129-169.

Kerr, J.F. (1963). *Practical Work in Science.* Leicester: Leicester University Press.

Khattri, N., Reeve, A.L. and Kane, M.B. (1998). *Principles and Practices of Performance Assessment.* New Jersey: Lawrence Erlbaum Associates.

Kreitler, H. and Kreitler S. (1974). The role of experiment in science education. *Instructional Science, 3,* 75-88.

Kreisberg, S. (1992). *Transforming Power: Domination, Empowerment and Education.* New York: State University of New York Press.

Lampert, M. (1985). How do teachers manage to teach?: perspectives on problems in practice. *Harvard Education Review, 55,* 178-94.

Lantz, O. and Kass, H. (1987). Chemistry teachers' functional paradigms. *Science Education 71(1),* 117-134.

Latham, H. (1877). *On the Action of Examinations Considered as a Means of Selection.* Cambridge: Deighton, Bell and Company.

Lau Chang, W.Y. (1980). *A Design for the Teacher Assessment of Practical Skills in Hong Kong Advanced Level Biology and a Study of its Criterion-related Validity.* Unpublished M.Ed. Dissertation, University of Hong Kong.

Layton, D. (1990). Student laboratory practice and the history and philosophy of science. In E. Hegarty-Hazel (ed.), *The Student Laboratory and the Science Curriculum,* pp.37-59. London: Routledge.

Lazarowitz, R. and Tamir, P. (1994). Research On Using Laboratory Instruction In Science. In D. Gabel (ed.), *Handbook of Research On Science Teaching And Learning,* pp.94-128. New York: Macmillan.

Lederman, N.G. (1992). Students' and Teachers' Conceptions of the Nature of Science: A review of the Research. *Journal of Research in Science Teaching, 39(4),* 331-359.

Lederman, N.G. (1999). Teachers' understanding of the nature of science and classroom practice: factors that facilitate or impede the relationship. *Journal of Research in Science Teaching, 36(8),* 916-929.

Lederman, N.G. and Zeidler, D.L. (1987). Science teachers' conceptions of the nature of science: Do they really influence teaching behaviour? *Science Education, 71(5),* 721-734.

Leinhardt, G. (1990). Capturing craft knowledge in teaching. *Educational Researcher, 19,* 18-25.

Lock, R.J. (1986). Assessment of Science Practical Skills in 15 year old pupils. *Unpublished PhD Thesis,* University of Leeds.

Lortie, D.C. (1975). *Schoolteacher: A Sociological Study.* Chicago: University of Chicago Press.

Louden, W. (1991). Understanding Teaching: continuity and change in teachers' knowledge. London: Cassell Educational.

Lyons, N. (1990). Dilemmas in knowing: Ethical and epistemological dimensions of teachers' work and development. *Harvard Educational Review, 60(2),* 159-80.

Madaus, G. (1993). A national testing system: Manna from above? A historical/technological perspective. *Educational Assessment, 1,* 9-26.

Marton, F. (1994). On the structure of teachers' awareness. In I. Carlgren, G. Handal, and S. Vaage (eds), *Teachers' Minds and Actions,* pp.29-42. London: Falmer Press.

Marton, F. and Booth, S. (1997). *Learning and Awareness.* New Jersey: Lawrence Erlbaum Associaties.

McLaughlin, M.W. (1987). Lessons from experience: Lessons from policy implementation. *Educational Evaluation and Policy Analysis. 9,* 171-178.

McLaughlin, M.W. (1990). The Rand Change Agent Study revisited: macro perspectives and micro realities. *Educational Researcher. 19(9),* 11-16.

McLaughlin, M.W. (1997). Rebuilding teacher professinalism in the United States. In A. Hargreaves and R. Evans (eds), *Beyond Educational Reform: Bringing Teachers Back In,* pp.77-93. Buckingham: Open University Press.

McLaughlin, M.W. and Talbert, J.E. (1993). *Teaching for Understanding: Challenges for Policy and Practice.* San Franscisco, Jossey-Bass.

McRobbie, C.J. and Tobin, K. (1995). Restrains to reform: The congruence of teacher and student actions in a chemistry classroom. *Journal of Research in Science Education, 32(4),* 373-385.

Millar, R. and Osborne, J. (1998). *Beyond 2000: science education for the future.* London: School of Education, King's College.

Mitchener, C.P. and Anderson, R.D. (1989). Teachers' perspective: developing and implementing an STS curriculum. *Journal of Research in Science Education, 26(4),* 351-369.

Morris, P. (1990). *Curriculum Development in Hong Kong. Education Paper No. 7.* Hong Kong: Faculty of Education, The University of Hong Kong.

Morris, P., Lo, M.L. and Adamson, B. (2000). Improving schools in Hong Kong: lessons from the past. In B. Adamson, T. Kwan and K. K. Chan (eds), *Changing the curriculum: the impact of reform on Hong Kong's primary schools,* pp.245-262. Hong Kong: Hong Kong University Press.

Munby, H. (1982). The place of teachers' beliefs in research on teacher thinking and decision making, and an alternative methodology. *Instructional Science, 11,* 201-225.

Murphy, R. and Torrance, H. (1988). *The Changing Face of Educational Assessment.* Milton Keynes: Open University Press.

National Science Council (1996). *National Science Education Standards.* Washington, D.C.: National Academy Press.

Nespor, J. (1987). The role of beliefs in the practice of teaching. *Journal of Curriculum Studies, 19,* 317-328.

Newton, P., Driver, R. and Osborne, J. (1999). The place of argumentation in the pedagogy of science. *International Journal of Science Education, 21,* 553-576.

Nott, M. (1997). Keeping scientists in their place. *School Science Review, 78*(285), 49-61.

Nyiri, J.C. (1988) Tradition and practical knowledge. In J. C. Nyiri and B. Smith (eds), *Practical Knowledge: Outline of a Theory of Traditions and Skills,* pp.17-52. London: Croom Helm.

Olson, J. (1980). Teacher constructs and curriculum change. *Journal of Curriculum Studies, 12(1),* 1-11.

Olson, J. (1981). Teacher influence in the classroom: a context for understanding curriculum translation. *Instructional Science, 10,* 259-275.

Olson, J., James, J. and Lang, M. (1999). Changing the subject: the challenge of innovation to teacher professionalism in OECD countries. *Journal of Curriculum Studies, 31(1),* 69-82.

Pang, K.C. (1992). The biology teacher assessment scheme (TAS). *Curriculum Forum, 2 (2),* 81-90.

Paechter, C. (1995). Subcultural retreat: negotiating the Design and Technology curriculum. *British Educational Research Journal, 21,* 75-87.

Pajares, M.F. (1992). Teachers' beliefs and educational research: cleaning up a messy construct. *Review of Educational Research, 62(3),* 307-332.

Pennycuick, P. (1990). Factors Influencing the Introduction of Continuous Assessment Systems in Developing Countries. In D. Layton (ed.), *Innovations in Science and Technology Education. Vol. III,* pp.139-52. UNESCO.

Peterson, J. (1992). *A Case Study of Evaluation of the Implementation of GCSE.* Unpublished PhD Thesis. University of Sussex.

Phillips, D.C. (1993). Telling it straight: issues in accessing narrative research. Quoted in Fenstermacher (1994) The knower and the known: the nature of knowledge in research on teaching. In L. Darling-Hammond (ed.), *Review of Research in Education 20,* 3-56.

Pong, W.Y. and Chow, J.C.S. (2002). On the pedagogy of examinations in Hong Kong. *Teaching and Teacher Education, 18,* 139-149.

Popham, W.J. (1987). The merits of measurement-driven instruction. *Phi Delta Kappa, 63,* 679-682.

Proudford, C. (1998). Implementing educational policy change: Implications for teacher professionalism and professional empowerment. *Asia-Pacific Journal of Teacher Education, 26(2),* 139-151.

Putnam, R.T. and H. Borko (1997). Teacher Learning: Implications of New Views of Cognition. In B. Biddle, T.L. Good and I.F. Goodson (eds), *International Handbook of Teachers and Teaching, Vol. II.* pp.1223-1296. Dordrecht: Kluwer Academic Publishers.

QSR NUD*DIST (1997). *QSR NUD*IST 4 User Guide.* London: Qualitative Solutions and Research.

Richardson, V. (1990). Significant and worthwhile change and teaching practice. *Educational Researcher, 19(7),* 10-18.

Richardson, V. (1996). The role of attitudes and beliefs in learning to teach. In J. Sikula, T.J. Buttery and E. Guyton (eds), *Handbook of Research on Teacher Education.* pp.102-119. New York: Macmillan.

Resnick, L.B. and Resnick, D.P. (1992). Assessing the thinking curriculum: New tools for educational reform. In B.R. Giggord and M.C. O'Conner (eds), *Changing assessments: Alternative views of aptitude, achievement and instruction,* pp.37-75. Boston: Kluwer.

Roehrig, G.H. and Luft, J.A. (2004). Constraints experienced by beginning secondary science teachers in implementing scientific inquiry lessons. *International Journal of Science Education, 26(1),* 3-24.

ROPES (1998). *Review of Public Examination System in Hong Hong. Interim Report.* Hong Kong: Hong Kong Examinations Authority, Hong Kong Baptist University.

Sadler, R. (1989). Formative assessment and the design of instructional systems. *Instructional Science, 18,* 119-144.

Sadler, R. (1998). Formative assessment: revisiting the territory. *Assessment in Education, 5(1),* 77-84.

Sanger, J. (1990). Awakening a scream of consciousness: the critical group in action research. *Theory into Practice, 29,* 174-178.

Shavelson, R. and Stern, P. (1981). Research on teachers' pedagogical thought, judgements, decisions, and behaviour. *Review of Educational Research, 51,* 455-498.

Shepard, L. (2000). The role of assessment in a learning culture. *Educational Researcher, 29(7),* 4-14.

Shulman, L. (1986). Those who understand: Knowledge growth in teaching. *Educational Researcher, 15*(2), 4-21.

Sikes, P., Measor, L. and Wood, P. (1985). *Teacher Careers: crises and continuities.* Lewes: Falmer Press.

Spillane, J.P. (1999). External reform initiatives and teachers' efforts to reconstruct their practice: the mediating role of teachers' zones of enactment. *Journal of Curriculum Studies, 31(2),* 143-175.

Spillane, J.P. and Jennings, N.E. (1997). Aligned instructional policy and ambitious pedagogy: exploring instructional reform from the classroom perspective. *Teachers College Record, 98(3),* 449-481.

Stigler, J.W. and Hiebert, J. (1999). The Teaching Gap: Best Ideas from the World's Teachers for Improving Education in the Classroom. New York: Free Press.

Synder, J., Bolin, F. and Zumwalt, K. (1992). Curriculum ImplementationCurriculum. In W. Jackson (ed), *Handbook of Research on Curriculum,* pp. 402-435. New York: Macmillan.

Thompson, J.J. (ed.) (1975). *Practical Work in Sixth Form Science.* Oxford: Department of Educational Studies, University of Oxford.

Tobin, K. (1986). Secondary science laboratory activities. *European Journal of Science Education, 8(2),* 199-211.

Tobin, K., Espinet, M., Byrd, S.E. and Adams, A. (1988). Alternative perspectives of effective science teaching. *Science Education, 72,* 433-451.

Tobin, K. and Fraser, B. (1989). Barriers to higher-level cognitive learning in high school science. *Science Education, 73,* 659-682.

Tobin, K. and Gallagher, J.J. (1987). What happens in high school science classrooms? *Journal of Curriculum Studies, 19,* 549-560.

Tobin, K. and Jakubowski, E. (1990). *Cognitive processes and teacher change.* Paper presented at the annual meeting of the American Educational Research Association, Boston.

Tobin, K., Kahle, J. and Fraser, B. (1990). Windows into Science Classrooms: Problems Associated with High-Level Cognitive Learning in Science. London: Falmer Press.

Tobin, K. and LaMaster, S. (1995). Relationships between metaphors, beliefs and actions in a context of science curriculum change. *Journal of Research in Science Teaching, 32(3),* 225-242.

Tobin, K. and McRobbie, C. (1996). Cultural myths as constraints to the enacted science curriculum. *Science Education, 80(2),* 223-241.

Tobin, K. and McRobbie, C. (1997). Beliefs about the nature of science and the enacted science curriculum. *Science and Education, 6,* 355-371.

Tobin, K., Tippins, D. J. and Hook, K.S. (1994). Referents for changing a science curriculum: A case study of one teacher's change in beliefs. *Science and Education, 3,* 245-264.

Todhunter, I. (1873). The Conflict of Studies and Other Essays on Subjects Connected with Education. London: Macmillan.

Torrance, H. (1991). Evaluating SATs – the 1990 pilot, *Cambridge Journal of Education, 21,* 129-140.

Torrance, H. (ed.) (1995a). *Evaluating Authentic Assessment.* Buckingham: Open University Press.

Torrance, H. (1995b). Teacher involvement in new approaches to assessment. In H. Torrance (ed.), *Evaluating Authentic Assessment,* pp.44-56. Buckingham: Open University Press.

Torrance, H. and Pryor, J. (1998). *Investigating Formative Assessment.* Buckingham: Open University Press.

Tunstall, P. and Gipps, C. (1996). Teacher feedback to young children in formative assessment: a typology. *British Educational Research Journal, 22,* 389-404.

Verjovsky, J. and Waldegg, G. (2005). Analyzing beliefs and practices of a Mexican high school biology teacher. *Journal of Research in Science Teaching, 42(4),* 465-491.

Wallace, J. and Louden, W. (2000). *Teachers' Learning: Stories of Science Education.* Dordrecht: Kluwer.

Watson, J.R., Swain, J.R.L. (2004). Students' discussions in practical scientific inquiries. *International Journal of Science Education, 26(1),* 25-45.

Wellington, J.J. (1989). *Skills and Processes in Science Education.* London: Routledge.

Wellington, J.J. (1998). Practical Work in School Science: Which Way Now? London: Routledge.

Woods, P., Jeffrey, B., Troman, G. and Boyle, M. (1997). *Restructuring Schools, Reconstructing Teachers: responding to change in the primary school.* Buckingham: Open University Press.

Woolnough, B. and Allsop, T. (1985). *Practical Work in Science.* Cambridge: Cambridge University Press.

Woolnough, B. and Toh, K.A. (1990). Alternative approaches to assessment of practical work in science. *School Science Review, 71(256),* 127-131.

Yip, D.Y. and Yung, B.H.W. (1998). Providing meaningful contexts for scientific investigations. *Australian Science Teachers' Journal, 44(1),* 35-42.

Yung, B.H.W. (1992). An Evaluation Study of the Feasibility of a Teacher Assessment Scheme in the Hong Kong Advanced Level Biology Examination. Unpublished M.Ed. dissertation, University of Hong Kong.

Yung, B.H.W. (1995). Fostering professional competence and development: Experience of the HKAL Biology Teacher Assessment Scheme. In F. Lopez-Real (ed.), *Teacher Education in the Asian Region. Proceedings of ITEC'95,* pp.400-408. Hong Kong: University of Hong Kong.

Yung, B.H.W. (1996). Can two variables be tested at the same time: A question failing not only pupils, but teachers too! In K. Volk (ed.), *Science & Technology Education Conference '98 Proceedings,* pp. 222-230. Hong Kong: Hong Kong Institute of Education.

Yung, B.H.W. (2001). Examiner, policeman or students' companion: teachers' perceptions of their role in an assessment reform. *Educational Review, 53(3),* 251-260.

Index

Science & Technology Education Library

Series editor: William W. Cobern, *Western Michigan University, Kalamazoo, U.S.A.*

Publications

1. W.-M. Roth: *Authentic School Science.* Knowing and Learning in Open-Inquiry Science Laboratories. 1995 ISBN 0-7923-3088-9; Pb: 0-7923-3307-1
2. L.H. Parker, L.J. Rennie and B.J. Fraser (eds.): *Gender, Science and Mathematics.* Shortening the Shadow. 1996 ISBN 0-7923-3535-X; Pb: 0-7923-3582-1
3. W.-M. Roth: *Designing Communities.* 1997 ISBN 0-7923-4703-X; Pb: 0-7923-4704-8
4. W.W. Cobern (ed.): *Socio-Cultural Perspectives on Science Education.* An International Dialogue. 1998 ISBN 0-7923-4987-3; Pb: 0-7923-4988-1
5. W.F. McComas (ed.): *The Nature of Science in Science Education.* Rationales and Strategies. 1998 ISBN 0-7923-5080-4
6. J. Gess-Newsome and N.C. Lederman (eds.): *Examining Pedagogical Content Knowledge.* The Construct and its Implications for Science Education. 1999 ISBN 0-7923-5903-8
7. J. Wallace and W. Louden: *Teacher's Learning.* Stories of Science Education. 2000 ISBN 0-7923-6259-4; Pb: 0-7923-6260-8
8. D. Shorrocks-Taylor and E.W. Jenkins (eds.): *Learning from Others.* International Comparisons in Education. 2000 ISBN 0-7923-6343-4
9. W.W. Cobern: *Everyday Thoughts about Nature.* A Worldview Investigation of Important Concepts Students Use to Make Sense of Nature with Specific Attention to Science. 2000 ISBN 0-7923-6344-2; Pb: 0-7923-6345-0
10. S.K. Abell (ed.): *Science Teacher Education.* An International Perspective. 2000 ISBN 0-7923-6455-4
11. K.M. Fisher, J.H. Wandersee and D.E. Moody: *Mapping Biology Knowledge.* 2000 ISBN 0-7923-6575-5
12. B. Bell and B. Cowie: *Formative Assessment and Science Education.* 2001 ISBN 0-7923-6768-5; Pb: 0-7923-6769-3
13. D.R. Lavoie and W.-M. Roth (eds.): *Models of Science Teacher Preparation.* Theory into Practice. 2001 ISBN 0-7923-7129-1
14. S.M. Stocklmayer, M.M. Gore and C. Bryant (eds.): *Science Communication in Theory and Practice.* 2001 ISBN 1-4020-0130-4; Pb: 1-4020-0131-2
15. V.J. Mayer (ed.): *Global Science Literacy.* 2002 ISBN 1-4020-0514-8
16. D. Psillos and H. Niedderer (eds.): *Teaching and Learning in the Science Laboratory.* 2002 ISBN 1-4020-1018-4
17. J.K. Gilbert, O. De Jong, R. Justi, D.F. Treagust and J.H. Van Driel (eds.): *Chemical Education: Towards Research-based Practice.* 2003 ISBN 1-4020-1112-1
18. A.E. Lawson: *The Neurological Basis of Learning, Development and Discovery.* Implications for Science and Mathematics Instruction. 2003 ISBN 1-4020-1180-6
19. D.L. Zeidler (ed.): *The Role of Moral Reasoning on Socioscientific Issues and Discourse in Scientific Education.* 2003 ISBN 1-4020-1411-2

Science & Technology Education Library

Series editor: William W. Cobern, *Western Michigan University, Kalamazoo, U.S.A.*

20. P.J. Fensham: *Defining an Identity. The Evolution of Science Education as a Field of Research.* 2003 ISBN 1-4020-1467-8
21. D. Geelan: *Weaving Narrative Nets to Capture Classrooms.* Multimethod Qualitative Approaches for Educational Research. 2003 ISBN 1-4020-1776-6; Pb: 1-4020-1468-7
22. A. Zohar: *Higher Order Thinking in Science Classrooms: Students' Learning and Teachers' Professional Development.* 2004 ISBN 1-4020-1852-5; Pb: 1-4020-1853-3
23. C.S. Wallace, B. Hand, V. Prain: *Writing and Learning in the Science Classroom.* 2004 ISBN 1-4020-2017-1
24. I.A. Halloun: *Modeling Theory in Science Education.* 2004 ISBN 1-4020-2139-9
25. L.B. Flick and N.G. Lederman (eds.): *Scientific Inquiry and the Nature of Science.* Implications for Teaching, Learning, and Teacher Education. 2004 ISBN 1-4020-2671-4
26. W.-M. Roth, L. Pozzer-Ardenghi and J.Y. Han: *Critical Graphicacy.* Understanding Visual Representation Practices in School Science. 2005 ISBN 1-4020-3375-3
27. M.J. de Vries: *Teaching about Technology.* An Introduction to the Philosophy of Technology for Non-philosophers. 2005 ISBN 1-4020-3409-1
28. R. Nola and G. Irzik: *Philosophy, Science, Education and Culture.* 2005 ISBN 1-4020-3769-4
29. S. Alsop (ed.): *Beyond Cartesian Dualism.* Encountering Affect in the Teaching and Learning of Science. 2005 ISBN 1-4020-3807-0
30. P.J. Aubusson, A.G. Harrison and S.M. Ritchie (eds.): *Metaphor and Analogy in Science Education.* 2006 ISBN 1-4020-3829-1
31. B.H.W. Yung: *Assessment Reform in Science.* Fariness and Fear. 2006 ISBN 1-4020-3374-5